MAKE TECHNOLOGY GREAT AGAIN

Scott Tilley

Make Technology Great Again

Cover design © Scott Tilley
Cover photograph (baseball cap) © ImagePixel/Shutterstock
Cover photograph (eagle) © omnimoney/Shutterstock
Cover photograph (buttons) © Plan-B/Shuttersock
Back photo (digital flag) © Bruce Rolff/Shutterstock

Published by CTS Press

CTS
Press
www.preciouspublishing.biz/ctspress

ISBN-13: 978-0-9996446-2-1

ISBN-13: 978-0-9996446-3-8 (ebook)

TABLE OF CONTENTS

DEDICATION

To my father. Gone, but not forgotten.

PREFACE

The election of Donald Trump as President of the United States in November 2016 caused great upheaval in the country. His campaign motto, "Make America Great Again," really resonated with a lot of people. It thought it was a very catchy phrase, and it inspired the title of this book.

Not too long ago, the average person could go about their daily lives without much interaction with computers and connected devices. They could drive to work, enjoy social interaction, and manage their household without having to worry about hackers and surveillance. Those days are gone.

Technology today has become exceedingly complicated. In many cases, the average person is baffled by the behavior of the many electronic gadgets in their lives. Computer viruses, data breaches, and genetic engineering now affect everyone, but few of us truly understand how these technologies work or what to do about them. Technology is still useful, but it could be improved. Technology needs to be made great again.

Looking back at 2017, we celebrated the 40th anniversary of Star Wars and the Voyager probes, net neutrality was repealed, and the latest iPhone was introduced with Face ID. But to me, three things really stood out: bitcoins, breaches, and busts.

Bitcoins are a form of cryptocurrency and just one implementation of blockchain, a technology for a global, distributed ledger system that many people think represents the future of commerce.

Data breaches were in the news all year. As more computing

moves into the cloud, the data follows it. Poor programming practices, pervasive human error, and sloppy organizational governance all contributed to this problem.

There's always been regrettable instances of poor behavior in the workplace, but 2017 was unique in the number of public busts that took place. Alleged perpetrators were outed through the power of social media, something that would not have been possible even a few years ago.

I can be reached via email at TechnologyToday@srtilley.com. You can follow my column @TechTodayColumn on Twitter. I'm also on Facebook: www.facebook.com/stilley.writer. Learn more about all of my writing at www.amazon.com/author/stilley.

Scott Tilley
Melbourne, FL
May 25, 2018

ACKNOWLEDGMENTS

I find inspiration for my columns in many places. Indeed, with technology changing so fast, I rarely lack for topics – I just lack time. Several organizations I'm involved with continue to provide a robust intellectual forum that I use to expand upon common technology themes. These are *Big Data Florida* (www.BigDataFlorida.org), the *Center for Technology & Society* (www.CTS.today), and the Space Coast chapter of the *International Council on Systems Engineering* (www.incose.org/scc). Thanks to everyone involved in these wonderful groups.

As always, thanks to my family and pets for keeping me honest – and for reminding me that there's more to life than work.

Lastly, thanks to all the loyal followers of my Technology Today column. I appreciate your correspondence and feedback. Without such great readers, this delightful compilation would not exist.

Calendar for Year 2017 (United States)

timeanddate.com

January

S	M	T	W	T	F	S
1	2	3	4	5	6	7
8	9	10	11	12	13	14
15	16	17	18	19	20	21
22	23	24	25	26	27	28
29	30	31				

◑:5 ○:12 ◐:19 ●:27

February

S	M	T	W	T	F	S
			1	2	3	4
5	6	7	8	9	10	11
12	13	14	15	16	17	18
19	20	21	22	23	24	25
26	27	28				

◑:3 ○:10 ◐:18 ●:26

March

S	M	T	W	T	F	S
			1	2	3	4
5	6	7	8	9	10	11
12	13	14	15	16	17	18
19	20	21	22	23	24	25
26	27	28	29	30	31	

◑:5 ○:12 ◐:20 ●:27

April

S	M	T	W	T	F	S
						1
2	3	4	5	6	7	8
9	10	11	12	13	14	15
16	17	18	19	20	21	22
23	24	25	26	27	28	29
30						

◑:3 ○:11 ◐:19 ●:26

May

S	M	T	W	T	F	S
	1	2	3	4	5	6
7	8	9	10	11	12	13
14	15	16	17	18	19	20
21	22	23	24	25	26	27
28	29	30	31			

◑:2 ○:10 ◐:18 ●:25

June

S	M	T	W	T	F	S
				1	2	3
4	5	6	7	8	9	10
11	12	13	14	15	16	17
18	19	20	21	22	23	24
25	26	27	28	29	30	

◑:1 ○:9 ◐:17 ●:23 ◑:30

July

S	M	T	W	T	F	S
						1
2	3	4	5	6	7	8
9	10	11	12	13	14	15
16	17	18	19	20	21	22
23	24	25	26	27	28	29
30	31					

○:9 ◐:16 ●:23 ◑:30

August

S	M	T	W	T	F	S
		1	2	3	4	5
6	7	8	9	10	11	12
13	14	15	16	17	18	19
20	21	22	23	24	25	26
27	28	29	30	31		

○:7 ◐:14 ●:21 ◑:29

September

S	M	T	W	T	F	S
					1	2
3	4	5	6	7	8	9
10	11	12	13	14	15	16
17	18	19	20	21	22	23
24	25	26	27	28	29	30

○:6 ◐:13 ●:20 ◑:27

October

S	M	T	W	T	F	S
1	2	3	4	5	6	7
8	9	10	11	12	13	14
15	16	17	18	19	20	21
22	23	24	25	26	27	28
29	30	31				

○:5 ◐:12 ●:19 ◑:27

November

S	M	T	W	T	F	S
			1	2	3	4
5	6	7	8	9	10	11
12	13	14	15	16	17	18
19	20	21	22	23	24	25
26	27	28	29	30		

○:4 ◐:10 ●:18 ◑:26

December

S	M	T	W	T	F	S
					1	2
3	4	5	6	7	8	9
10	11	12	13	14	15	16
17	18	19	20	21	22	23
24	25	26	27	28	29	30
31						

○:3 ◐:10 ●:18 ◑:26

Jan 1	New Year's Day	May 14	Mother's Day	Nov 11	Veterans Day
Jan 2	'New Year's Day' observed	May 29	Memorial Day	Nov 23	Thanksgiving Day
Jan 16	Martin Luther King Day	Jun 18	Father's Day	Dec 24	Christmas Eve
Feb 14	Valentine's Day	Jul 4	Independence Day	Dec 25	Christmas Day
Feb 20	Presidents' Day	Sep 4	Labor Day	Dec 31	New Year's Eve
Apr 13	Thomas Jefferson's Birthday	Oct 9	Columbus Day		
Apr 16	Easter Sunday	Oct 31	Halloween		

LOOKING AHEAD TO 2017

Drone delivery, space program, technology & society

January 6, 2017

Each January I join other pundits in predicting the future of technology for the coming year. My track record is mixed, but it's always fun to do. This year, I thought I'd describe three developments that have varying chances of becoming reality. One that I think will happen (drone delivery), one that might happen (a renewed space program), and one that should happen (society takes more interest in technology).

Drone delivery: On December 7, 2016, Amazon.com made its first commercial drone delivery. A drone delivered a small package to a rural home near Cambridge, England. It took just 13 minutes from order placement to drone fulfillment. A modest first step, to be sure, but it did mark an important milestone in Amazon.com's goal to begin using drone fleets for widespread product transportation. Amazon Prime Air is no longer fiction; it is real.

There are still significant technical hurdles to make drone delivery commonplace, but perhaps the biggest challenge is regulatory. There's a reason the maiden flight was in the UK and not the USA. The FAA is still struggling with how to provide oversight of drone traffic, both commercial and private, and until things get better, we'll be behind the curve of this exciting new innovation.

Space program: There is a lot of hope within the space community that the incoming administration will increase funding and possibly renew NASA's original mission of getting us to the stars. Recent years witnessed a series of project cancelations that had significant negative impact our local economy. The decisions also affected our standing as a global leader in space exploration.

Our own Buzz Aldrin envisions a new commercial "race for space" in 2017. Players such as SpaceX and Blue Origin have already made their presence felt on the Space Coast. I'm cautiously optimistic that positive changes may happen for our national space program. We have the people and we're developing the technology. We just need the proper policies in place to make it happen.

Technology & society: I closed out 2016 with three surreal topics that were in the news last year: artificial intelligence (AI), big data, and cybersecurity. These are important technological developments that have the potential to dramatically affect society at-large. But they're also technologies that few people understand – and that's a problem.

I founded the Center for Technology & Society to address this very issue. I firmly believe that everyone should become better informed about how advances in STEM will change their lives, both positively and negatively. I think an educated population is key to an advanced modern society, and we all have the ability to learn.

# # #

THE HUMAN FACE OF BIG DATA

Using big data analytics to uncover hidden patterns in life

January 13, 2017

This week, the Big Data Florida user group kicked off 2017 with the captivating documentary "The Human Face of Big Data." The movie (and book of the same name) are products of "a globally crowd-sourced media project focusing on humanity's new ability to collect, analyze, triangulate, and visualize vast amounts of data in real time," created by Rick Smolan and Jennifer Erwitt. It's a fascinating account of the impacts of big data on society. You can find a trailer for the documentary online at http://bit.ly/1HuByPc.

After watching the film, we had a facilitated discussion about some of the main points raised by the authors and the experts they interviewed, including Esther Dyson and Mark Goodman. There were a number of themes that emerged, including the role big data already is playing in areas such as health care, disaster response, and the judicial system.

Scientific advancements have often been driven by innovative use of new technology. Galileo used the telescope in 1610 to discover four moons of Jupiter and previously unseen features of our Moon. Over three hundred years later the electron microscope was used to see details of incredibly tiny objects. Now, another hundred years have passed and we have the first uses of big data analytics to uncover patterns previously hidden to us. In each case, the data (moons, particles, or relationships) already existed; we just lacked the tools to see them. The difference with big data is that the patterns are in the virtual world, not the physical world.

One of the key elements making big data more accessible to business is the power of visualization. Pictures are literally worth a thousand words – or more. Just ask Edward Tufte, a pioneer in the visual display of quantitative information.

It was interesting to see several new terms associated with big data be introduced during the interviews. For example, someone from the New York Times has a formal position as "data artist" for the newspaper. A researcher from MIT who recorded all of the formative years of his young child's daily activities referred to the child's first attempt at speaking as "word birth" and the gradual use of these words in sentence fragments as "wordscapes." Several experts had the on-screen title as "futurist."

This meeting provided a good overview for those just getting started in big data – and a great refresher for everyone else clearing out the holiday cobwebs. It also provided a solid framework for our next three meetings this Spring, which are on big data and the Internet-of-Things (IoT), machine learning and robotics, and data science and sustainability. Feel free to join us!

#

THE SORRY STATE OF SOFTWARE

After all these years, crashes are inexcusable

January 20, 2017

If I see one more spinning beach ball in the web browser or experience total system lockup on my notebook computer one more time, I'm going to completely lose it. As someone who has worked in the technology field for literally decades, I'm ashamed to say that the sorry state of software today is inexcusable.

It's time to shift into full curmudgeon mode.

I've been using web browsers since Mosaic made its first appearance in January 1993 – 24 years ago this month. My primary browser, Safari, was released in January 2003 – 14 years ago this month. Mosaic may be long gone, but Safari is still widely used. And it still hangs and crashes all the time. After being developed, optimized, and (presumably) tested, there's just no legitimate excuse for such a vital piece of software to act like this. None. Such buggy behavior wouldn't be accepted in other products, like automobiles (they'd be recalled), so why do we put up with this deplorable situation for software applications?

This week I was trying to do some research for an ongoing project. In today's world, the word "research" is basically a synonym for "Google," so I had multiple tabs open in the browser, each loaded with websites, PDF files, or videos. Quite soon the browser became unresponsive, leaving me frustrated while I waited for the beach ball to complete its annoying dance on the screen. And of course this always happens when you're in a hurry, such as trying to print a document for a meeting.

The lack of visibility into the underlying processes on the computer is particularly vexing. It's very difficult to know exactly what's going on. The browser could be hung for any number of reasons: a slow website, malware, Flash getting ready to crash (again), DNS servers slumbering somewhere along the network, and so on. But the average user has no way of knowing any of this; the system has just stopped working. Sometimes it will come back to life, but you never know how long to wait. Other times, a hard reboot is needed – both figuratively and literally.

Programs like Microsoft Word, which was first released way back in 1983, have even less excuse for their poor behavior. I know there are alternatives, but realistically, I'm stuck with it. It makes me cringe when I think of the lost productivity across the country. I've lost track of the number of times I get a message about Word crashing and losing my writing. "Would I like to send a report?" You're darn right I would! But I doubt it would be suitable for family viewing.

# # #

WHY PROGRAMMING PROBLEMS PERSIST

We need to stop accepting "good enough" software

January 27, 2017

Last week I wrote about the sorry state of software. From all the feedback I received, it seems I'm not the only one frustrated with spinning beach balls and crashing applications. Just today, the following error message was displayed when a video could not play in the browser: "Error: undefined is not an object (evaluating 'value.autoplay')."

Ignoring for now the fact that the error message is not even a proper sentence, whom is it meant for? The average user? Obviously not. It's more of a cryptic message in a bottle, written by and for the same person: the developer who wrote the code. And therein lies the heart of why programming problems persist.

The quality of software today is generally poor because it can be. We as consumers accept "good enough" applications because we don't know any better. The general public views programming as a black art, ignorant of what programmers actually do. Saying "sling Python code" is equivalent to saying "add eye of newt" – both are incomprehensible phrases only wizards understand. That must change. Everyone needs to improve their computing literacy; it's just as important as regular literacy and numeracy.

People assume programmers are doing their best while working in a very complex environment. They're not. At least, not when it comes to your interests. It's commonplace for programmers to knowingly ship code with bugs (defects) because they are under intense pressure to get it done as fast as possible. Testing is supposed to catch the errors before they make it to the customer, and it does – but not completely.

The reality is, the number one requirement in most software projects is to ship it fast. Quality is a long way down the list. Companies may tell you otherwise, but except for safety-critical systems, that's the dirty truth. The assumption is that the application can be patched later, after it's shipped. It's a terribly lazy way of doing things.

A complicating factor is that programming is an activity unlike any other: we actually celebrate amateurs. There's a perception that software engineering is just programming. But taking a two-week online course in React and Node.js doesn't make you an engineer – it makes you a hacker. The result is a lot of self-taught developers who know just enough to be dangerous working in the field. This is not healthy.

Would you go to a surgeon who was self-taught? Or a lawyer who never passed the bar or even obtained a degree, but learned everything from watching "Law & Order" episodes on TV? No. But we elevate such people in programming to skilled professionals. They are not.

# # #

HOW TO FORCE FIXES ON A FAULTY SYSTEM

A carrot and stick approach

February 3, 2017

Another week, another series of software glitches. This time it was Delta Air Lines who canceled nearly 300 flights due to a "computer failure." Last week it was United's turn to cancel flights due to unspecified "IT issues."

Ground stops like this are far more serious than browser bugs, but in many cases the root cause is the same: low quality software. I wrote about some of the reasons for this sorry state of affairs last week. Since things don't seem to be getting any better, I believe it's up to us – the consumers – to force changes upon this faulty system.

In some professions, such as medicine, government provides regulatory oversight. Doctors can be banned from practice if they commit egregious errors. Patients can also sue them if something goes horribly wrong with a procedure.

These are not options in the software world because programmers, testers, and others involved in crafting applications are not professionals in the same way. They are not licensed. They may be culpable for errors in the code, but they are not liable for them. This will only change if and when we demand it; there is currently no incentive for the industry to self-regulate their actions.

There are definitely a lot of downsides to installing the equivalent of the EPA or the NRC to manage the development of software. Bureaucratic overreach would inevitably lead to administrative bloat that still might not fix the problems. Modern society relies on creativity and out-of-the-box thinking to invent new technologies – traits not commonly associated with government committees.

If lawsuits and regulations are the stick wielded against the industry as a whole, a carrot might work better on individuals. There needs to be a significant change of mindset when it comes to the people who are involved in the engineering process. If they adopted a more user-focused and customer-centric approach to everything they do, they might find that instead of dealing with angry users they'd be receiving compliments on their work.

The education and training that is offered to programmers and testers clearly has room for improvement too. We've been developing software for over 50 years. You might think that key lessons learned from this experience would be codified by experts and internalized by neophytes when they enter the field. To some extent, they do, but there are many instances where they don't. How can browsers cause a complete system shutdown when we've been developing them for over 20 years? What happened to the corporate knowledge?

There's plenty of opportunity to make new mistakes. We don't need to keep making the same old ones again and again.

# # #

TRANSPORTATION SECURITY

Managing air, rail, sea, and even space traffic is essential

February 10, 2017

A recent trip home from Montreal to Melbourne took me nearly 30 hours. It should have taken me 8. The cascade of mechanical problems, poor customer service, and overall incompetence left me tired and frustrated. I lost time. I lost sleep. But at least I eventually made it home safe and sound.

The whole experience made me realize how susceptible our air travel system is to a single point of failure. Just one thing going wrong causes a terrible domino effect. Unfortunately, I can only see the situation getting worse as traffic levels increase.

Air travel is just one form of transportation that makes up our national infrastructure. Consider cargo traffic, which has increased significantly in the last few years. Cargo ships have become gargantuan platforms that carry huge loads across the oceans. Ports around the world are constantly being re-dredged to accommodate these floating behemoths. One of the biggest cargo ships in the world, the CSCL Globe, is more than four football fields long. It can carry 19,000 twenty-foot containers. Think how many 18-wheel transport trucks that means on the highways.

How do we know what's inside each of these cargo containers? What technology do we use to ensure that weapons are not smuggled into the country? Once the containers are unloaded from the ship, what rail and road routes do they take before they reach their final destination?

The volume of trucks and cars on our roads is also growing. In many parts of the world, the rising middle class is resulting in a surge of highway traffic. In 2010, there was a traffic jam outside of Beijing, China that lasted for almost two weeks. Nearly 20 lanes of traffic stretched for more than 60 miles.

And you thought your commute was bad.

The amount of time people waste in their car, stuck in traffic during their daily commutes, continues to increase. In some big cities, spending more than four hours a day – each way – has become the norm. Not only is this terribly stressful on the driver (and passengers), it's a colossal loss of productivity. Our national GDP suffers from gridlock. It's also a mounting security risk.

Tonight at 8:00 p.m. in the Henegar Center, Dr. Cliff Bragdon will be speaking about transportation security as part of the Center for Technology & Society's "Tech Talk" series. Tickets are just $10 and can be ordered online at www.henegar.org or by calling the box office at (321) 723-8698. I hope you come to hear about some of the many challenges facing our national transportation system – and some of the possible solutions to avoiding intermodal gridlock in the future.

# # #

The Internet-of-Things (IoT)

The next big thing is already here

February 17, 2017

The "Internet of Things" (IoT) refers to a growing collection of connected devices. The "things" can be almost anything that can be placed on a network: appliances, automobiles, computers, smartphones, home automation systems, health sensors, even pet identification chips. If the device can go online (e.g., using a Wi-Fi signal), then it can be connected to the Internet.

This explosion in connected devices is already happening. I use a Nest thermostat that I can control remotely. I use Amazon.com's Echo to talk with Alexa in the cloud. I wear a FitBit to track my daily activities (too low) and heart rate (too high). I even use little Dash buttons scattered around my house to quickly reorder supplies from Amazon.com Prime.

I choose to use these devices because their benefits currently outweigh the risks, but that may change over time. Every connected device is a cybersecurity problem just waiting to happen. Last year there was a massive online attack that affected huge swathes of the Internet by commandeering IoT devices such as baby cameras and home routers. Most people don't update the software on these devices, so once a vulnerability is known, it never goes away. It's like leaving your front door open all the time.

At the Big Data Florida meeting this week, Karl Seiler spoke about some of the warnings related to IoT, which included cybersecurity. But there are other concerns as well, such as fake data.

Just as "fake news" has been in the news recently, there is a possibility for hackers to inject fake data into the network. If a control system was relying on data from sensors, this could be exploited to make the system respond inappropriately. For example, a reservoir could release a cascade of water if it was told overflow conditions were present, even if there was a drought. Think about the recent issues with the crumbling dam in California and massive evacuation that ensued.

IoT may usher in a new era of automation. With millions of inexpensive sensors deployed in domains such as transportation, driverless cars (and trucks) will be much easier to manage on the roadways. But this would mean fewer human drivers, fewer gas stations, and less fleet maintenance. Whether or not this is a good thing or a bad thing depends on which side of the automation equation you stand: the side that benefits from reduced cost, or the side that loses their job due to robots and artificial intelligence driven by IoT data streams.

Look around your home or office. Imagine literally everything you see connected to the Internet. That's the IoT future speeding your way.

#

ROBOT ENGINEERS WEEK

The profession evolves in unexpected ways

February 24, 2017

This is National Engineers Week (EWeek). According to the National Society of Professional Engineers, "EWeek is dedicated to ensuring a diverse and well-educated future engineering workforce by increasing understanding of and interest in engineering and technology careers." I think very soon we'll acknowledge, if not celebrate, a special type of engineer: the robot engineer.

You might think a robot engineer is a person who builds robots. But I'm referring to a robot that does engineering. We already have robots that perform sophisticated manufacturing tasks, so taking the next logical step and adding programmed creativity is not too far-fetched.

Consider software engineering. Last month I railed about the sorry state of most software applications: buggy, with idiosyncratic user interfaces, and unreliable functionality. The root cause of this low quality rests with the people who develop the applications. No matter how good their intentions, they are human and therefore prone to failure. Robots, on the other hand, rarely fail, and certainly not in the same way humans do.

We've had the ability for one program to write another program for decades. For example, I used to rely on programs to generate scanners and parsers used in source code compilers. As soon as a problem can be unambiguously described, it can be automated. Much of programming involves understanding the problem, identifying the best solution, and then crafting the code that implements it. Developers reuse code all the time to make their jobs easier. Robot engineers would just take this to the next level.

Employers in high-tech areas are constantly searching for new engineering talent. In many cases, corporate growth is stymied by a lack of suitable human resources. What if that problem went away? What if there was a way to hire (and program) a fleet of robots that could carry out many (if not all) of the engineering duties needed? Companies would be all over this as a way of saving money and improving quality.

Is it really a stretch to think that an electrical engineer who designs circuit boards could be replaced with a robot that had been given the same knowledge? Or a civil engineer involved in city planning?

I can already hear the neo-luddites decrying the introduction of robots to the engineering profession, but it's inevitable. Maybe not today or tomorrow, but it will happen. A few years ago, Scott Adams introduced a robot character in his comic strip to work alongside engineers Dilbert, Wally, and Alice, and in this case, life imitates art.

Robot engineers will certainly help with the "diversity" goal of EWeek, although perhaps not in the way it was originally intended.

# # #

CHATBOTS

What goes around, comes around

March 3, 2017

One of the modern scourges of so-called customer service centers is the call tree. These are the automated systems that require you to enter '1' to continue the call in Klingon, '2' to leave a message that will never be answered, or '3' to be placed in a seemingly infinite queue with elevator music playing in the background – with a few advertisements slipped into the stream. While you wait, you are assured that "your call is important to us … which is why we're not answering it." If someone does eventually answer, they inevitably ask you for the exact same information that you laboriously entered on the telephone keypad. Then they'll hang up.

Many of these call centers are operated by firms overseas. Over the last decade or so, companies outsourced their customer service functions in the hopes of reducing costs. These costs are primarily the salaries and benefits of people who work the phones here at home. Costs for representatives in the developing world were much less, and the feeling was that customers who are calling for help wouldn't notice the difference, or wouldn't care.

They did, and they do.

There's been a political backlash against outsourcing jobs of all kinds lately, particularly in the manufacturing sector, and these phone centers are no different. Some companies have brought their customer service back to the mainland, "re-shoring" the work with local people. But most other companies are gearing up for another solution: robots.

These are not physical robots, but intelligent programs that can mimic much of the functionality of an actual customer service representative. These programs are called chatbots, and they're a very hot topic these days. Companies like Facebook and Microsoft are pouring huge amounts of research and development dollars into developing chatbots that other companies can use.

We're already accustomed to interacting with chatbots: Siri and Alexa are good voice-based examples. Other chatbots rely on text messages sent through apps like Facebook Messenger or WhatsApp to carry on conversations. The intention is that the customer never realizes the "person" on the other end is not a person at all, but a program. This has been a goal in the artificial intelligence community for many years, going back to expert systems such as Eliza from MIT in the 1960s.

A recent study suggested that for every manufacturing job lost to overseas competitors, eight jobs are lost to robots and automation. The irony here is that this time the jobs are leaving the developing world and going into the virtual world. I guess what goes around, comes around, but the societal upheaval from such technological disruptions is evitable.

# # #

THE MATRIX REVISITED

Are today's robots the progenitors of tomorrow's Agents?

March 10, 2017

Each Spring Break I watch a TV series or a movie collection with a shared theme. I've been thinking a lot lately about dystopian futures where robots rule the world, so this year I decided to revisit "The Matrix." It's been an interesting experience to go back down this digital rabbit hole after so many years.

The first movie came out in 1999. The two sequels, "Reloaded" and "Revolutions," both came out in 2003. The franchise has earned more than $1.6 billion worldwide since its initial release, making it one of the most successful post-apocalyptic cyberpunk series ever.

The story is set a little over 100 years in the future. The human race has been reduced to living deep underground while they battle the robots that rule the surface. Ironically, it was the humans who created the robots and imbued them with artificial intelligence. The machines quickly became sentient and enslaved their makers.

The matrix is a virtual reality environment that most of the population inhabits without even knowing it. Humans have become the primary power source for the machines. Each person is grown in a pod, connected to the grid, oblivious to the real world. They think they are living normal lives in a typical urban environment, but the truth is they are nothing more than organic batteries for the robot civilization.

A few people are able to break free from the matrix and escape to their ruined world, but they are constantly immersing themselves in the artificial world to rescue more humans. While in the matrix, humans battle programs called Agents. One such program, Agent Smith, goes rogue and eventually enters the real world, setting up a final conflict between the machines and Neo, the main character with unique abilities to actually see the source code underlying the matrix.

Stylistically, "The Matrix" is a unique film. The special effects were groundbreaking when the first movie came out and they still hold up well today. The writers foresaw swarms of robots, called Sentinels, which attack the human's ships like an advanced army of autonomous drones.

Stories about man-made disasters and failed societies are nothing new. Whether the destruction comes about because we've engineered our own killers as in "The Matrix," or the world becomes populated with infected zombies as in "The Walking Dead" or "Z Nation," the result is the same: societal collapse.

Could something like "The Matrix" really happen to us? Are we the creators of our own imminent demise? Come to the Henegar Center this Saturday at 2:00 p.m. to hear more about the role robots and artificial intelligence will have on our society.

# # #

APPLE WATCH

Retro wristwatches are not the future

March 17, 2017

"Apple Watch: everyone I know bought one, almost no one I know wears one. It's confusing and not very useful."

That quote is from Jason Calacanis, a serial entrepreneur and venture capitalist based in California. But I could just as easily have written it – except that very few people I know bought one.

For the past two weeks I've been wearing an Apple Watch Series 2. It's the first time I've worn a watch for well over 20 years, which may have been part of the problem. I just couldn't get used to the feeling of a big, heavy piece of electronics on my wrist all the time. I've always relied on my phone to see the time – a simple system that works fine. Besides, here in Florida, I found my wrist sweating under the watch, particularly with the active sensors that rest on your skin.

The Apple Watch is elegant looking. The display is crisp and clear. There are several nice bands you can choose from. But tiny the watch is not. It's actually quite bulky. I found myself hitting the watch face on objects without realizing it.

The watch has an innovative charging system that I wish the iPhone would adopt. The charging process was quite rapid. But the watch itself did not retain a charge for very long; I found it needed charging about once a day.

You can make Dick Tracy style phone calls (audio only) directly from the watch, which definitely ups the cool factor. You can also send and receive text messages and receive notifications on it. But these features require that your iPhone be close by, because the watch is basically tethered to it for these functions to work.

Many people buy the Apple Watch for its health and exercise capabilities. For example, it can track your steps, your pulse rate (but not your blood pressure), and your walking distance. It nags you to "get moving" at various parts of the day when you are more sedentary. But the showstopper for me was the watch's inability to track your sleep patterns. This is something that competing devices like the FitBit do very well.

I found that many of the apps on the phone had quite limited capability. The user interface is very different than the phone, due primarily to the much smaller screen size.

The bottom line is that the Apple Watch is a solution in search of a problem. It's just not really needed. The version I bought cost $399, so it's not a cheap trinket either.

I think implantable devices may have more of a future than retro wristwatches.

# # #

FOSSIL FUELS

Cheap, powerful, and dirty

March 24, 2017

Do you remember these lyrics?

"And then one day he was shootin' at some food, // And up through the ground come a bubblin' crude. // Oil that is, black gold, Texas tea."

Jed Clampett became a very rich man when he struck oil. But it wasn't just the Beverly Hillbillies that made their fortune from oil. One of our earliest business magnates, John Rockefeller, founded Standard Oil in 1870 and went on to become one of the richest men in the world. He had a product that we just couldn't get enough of – and our appetite for oil has only increased in the last 150 years.

According to the US Energy Information Administration (EIA), in 2015 the United States used over 7 billion barrels of petroleum products, such as gasoline, diesel fuel, and jet fuel. That's about 20 million barrels a day.

Having a steady source of oil has been an essential part of government policy for a long time. Over a hundred years ago, the British Royal Navy made the decision to switch from coal to oil for their fleet. We've been managing our petroleum sources on a geopolitical scale ever since. Today, our modern economy would literally grind to a halt without oil, which is why we do business with countries that don't really like us – but they like our money.

Until recently, the United States imported the majority of its oil.

As of 2015, the number one source of our oil was Canada, which surprises many people. Most people assume it's Saudi Arabia or some other part of the Middle East. Our number three source is Mexico, which means we're conveniently sandwiched between our major suppliers.

That all changed in the last few years with the introduction of fracking. Hydraulic fracturing involves the use of liquids injected at high pressure into subterranean structures, which forces open the ground and enables the extraction of oil and gas. From a purely technical perspective, fracking is an incredible advance in petroleum engineering. From an ecological perspective, fracking shares some of the lamentable characteristics of most oil exploration methods: pollution and environmental damage.

Fracking has put the United States on the road to energy independence. It's a remarkable change from just a few years ago, when industry analysts were predicting "peak oil" and the inevitable exhaustion of petroleum sources. Fracking has changed the industry so much that a gallon of gasoline is literally cheaper than a gallon of bottled water.

Fossil fuels are cheap, powerful, and they make some people very rich. To replace them, any new source of energy will have to address all three issues.

# # #

RENEWABLE ENERGY

Hydroelectric, wind, and solar provide a fraction of our needs

March 31, 2017

Contrary to popular belief, fossil fuels are renewable sources of energy. The problem is the timescale needed for the renewal process is millions of years. We use fossil fuels much faster than they can be replaced, which is one of the reasons why there's an ongoing search for other sources of energy that are renewable on a human timescale.

The three main sources of renewable energy that provide power to the country's electrical grid are hydroelectric, wind, and solar. As of 2015, these three sources contributed about 15% towards our energy needs. Hydroelectric power makes up about half that, with wind about a third and solar about 5 percent.

Renewable energy has the obvious benefit that it is naturally replenished. For example, we're not going to run out of solar power as long as the Sun shines in the sky. The more renewable energy we're able to generate, the less we'll have to rely on fossil fuels, which generally means less harm to the environment. This could improve our chances of managing climate change and the problems associated with it.

However, renewable energy is not without problems. Consider hydroelectric power, which is created when fast-moving water is harnessed to spin turbines that generate electricity. Building hydroelectric power stations often involves massive engineering projects that dam rivers and flood large tracts of land. The Three Gorges Dam in China produces enormous amounts of power, but construction of the dam displaced over a million people and caused significant ecological and cultural damage.

Wind is an unpredictable source of power: when it blows, wind turbines produce electricity; when it doesn't, they don't. Nevertheless, wind farms are popping up all over the country. People who live near them complain of the terrible low-frequency thrumming noise they produce. The turbine blades are huge and turn very fast, which has the unfortunate side effect of killing hundreds of thousands of bats and birds each year, including thousands of federally protected raptors such as bald and golden eagles. It's rather ironic that our national symbol is butchered to power our electric toothbrushes.

Solar power is the most passive renewable energy source: collectors gather sunlight and convert it to electricity. Of course, solar panels are not very efficient when the sun is not out. Some of the large solar arrays in the southwest desert are truly impressive, but in urban environments they're certainly not pretty to look at. Smart materials may solve this problem by embedding solar panels into other things, such as concrete or paint.

As the saying goes, nothing in life is free, and renewable energy is no different. We need to do better.

###

POWERING TOMORROW

Harnessing the real power of the sun

April 7, 2017

For over 150 years, the world has run on fossil fuels. For the last few decades, there has been a concerted effort to wean us from this unsustainable source of energy to renewable sources, such as hydroelectric, solar, and wind. After a lot of effort and a significant amount of capital investment, renewables provide just 15% of our total energy needs. Clearly we need a better solution.

The power sources of tomorrow will likely be a mix of our current sources, made better in terms of increased availability and decreased environmental impact, and new sources yet to be developed. I don't know what those new power sources will be, but I have confidence in our ability to create and/or harness them for the greater good.

One of the more promising sources of energy that we've yet to master is the type of power that makes life on Earth possible: nuclear fusion. We harness solar radiation from the Sun and convert it to electricity, but it's fusion that produces the radiation in the first place. There have been research projects underway for some time focused on fusion reactors, but they've proven to be devilishly difficult to master. However, if you can imagine a "Mr. Fusion" powering every car and house in the future, as it did in the movie "Back to the Future: Part II," then our energy problems would be all but solved.

Without nuclear fusion, we're left with nuclear fission, which is the process currently used in today's nuclear reactors. It's a very powerful source of energy, but socially unpopular due to its detrimental by-products. Since 9/11, nuclear reactors have also become security concerns. Three Mile Island, Chernobyl, and Fukushima are synonymous with meltdowns and nuclear fallout. From a purely political point of view, it's doubtful that this source of energy will experience any sort of renaissance in the near future – unless a new technology can be developed to provably secure the facility and manage the radioactive waste they produce.

There are many other, more esoteric, proposals for new sources of energy. Beaming power via laser or microwaves from space-based platforms. Matter/anti-matter collisions straight from the "Star Trek" playbook. Worm holes to draw power from distant black holes. I doubt any of these ideas will see the light of day, but that doesn't mean we shouldn't keep looking.

The reality is that our energy needs are growing each year, and as more parts of the world lift their populations out of poverty and into the middle class, their energy needs will only increase. Something needs to turn on the lights for 7.5 billion people.

###

WORLD WAR I

Lest we forget murderous technology

April 14, 2017

The reported use of Sarin gas in Syria recently caused a justifiable uproar in the worldwide community. Even war has rules, and using chemical weapons crosses a line set by the major powers in 1925, shortly after the end of World War I. The ban was put in place in part because of the horrific experiences of chlorine and mustard gas on the battlefield over a century ago.

Ironically, it was April 6, 1917 when the United States officially entered the First World War. 100 years later, almost to the day, the United States entered a new stage of the Syrian conflict by launching a cruise missile attack on Syria for its use of chemical weapons.

Global conflict is often a catalyst for technical innovation. World War I began in 1914 with a 19th century view of battle, with horses and cavalry. The war ended in 1917 with armored tanks, flamethrowers, machine guns, tracer bullets, poison gas, and depth charges.

As horrific as the use of World War I technology was, it was no match for what was to come two decades later. With World War II came the most destructive force ever created by mankind: nuclear weapons.

I recently re-watched the movie "The Day After," which was first broadcast in November 1983. It tells the story of a massive nuclear attack on the homeland, using Lawrence, Kansas as an example of what could happen when thermonuclear bombs are detonated in the atmosphere and near the ground. The result was physical devastation and societal collapse. The movie had a profound impact on the national psyche – and on me personally.

To better understand the scale of destruction possible by modern atomic bombs, consider Little Boy, which was dropped on the Japanese city of Hiroshima in 1945. It had the explosive power of about 15,000 tons of TNT. Sixteen years later, the Soviet Union exploded a bomb with a 50-megaton blast. That's over 3,000 times more powerful. I've little doubt that advances in technology in the intervening 50 years have only increased the yield of these devices, made even more powerful with multiple warheads per missile.

In the last 25 years, it seems we've moved on from worrying about nuclear annihilation to worrying about deadly viruses and rampaging zombies. Biological threats, such as the use of weaponized anthrax or Ebola, remain concerns. Cybersecurity is more in the news because it affects us all almost every day, albeit with far less consequences than being nuked. But the renewed use of chemical weapons makes me think that the meaning of the phrase from World War I, "Lest We Forget," has been forgotten.

# # #

AUGMENTED REALITY

Dive right in!

April 21, 2017

We've all seen it: crowds of people walking along busy streets, heads down, staring dumbly at their smartphones, seemingly oblivious to the world around them. The same scene plays out at restaurants everyday: everyone focused on their phone, not their dining companions.

If Facebook gets their way, it's only going to get worse.

At their recent developer conference called F8 (not to be confused with the newest "Fast and Furious" movie starring Dom and his buddies), Facebook unveiled an increased emphasis for augmented reality for their mobile apps. Think "Pokémon Go" on steroids.

With augmented reality, the user still sees and interacts with the real world. However, their view of the world is enhanced with virtual characters and images that overlay this view. For example, if you were looking at live video of your garden using your phone's built-in camera, your view of the garden could be augmented with virtual images of fluttering butterflies, waving gnomes, even fire-breathing dragons.

With Facebook's vision of the future, your smartphone's camera (and screen) becomes the most important part of the device. But you don't need any special equipment to experience this augmented reality: the computer in the phone blends the real and the virtual together into one seamless experience. A blank wall could become an interactive art canvas, complete with artist commentary.

This is quite different from virtual reality, which typically requires users to don bulky wrap-around glasses, like Facebook's own Oculus Rift headgear. Virtual reality is meant to be totally immersive: you are in an artificial world, completely created by the computer, with little sensory input from the real world.

Virtual reality has been around longer than augmented reality, but it's yet to become mainstream. There are a number of technical challenges still to overcome, such as minimizing lag times between user movement and system response. Even small delays can cause people to feel nauseous. The equipment itself tends to be expensive, with the exception of Google Cardboard, a simple virtual reality system that relies on your smartphone and a cardboard holder to create the simulated environment.

There's also a social factor that limits current virtual reality technology: absolutely no one looks good when wearing massive wrap-around glasses. Most photos of people using virtual reality headsets show them looking like dorks, with gaping mouths, seemingly staring into space, oblivious to the real world. But that's the whole point of virtual reality.

We already accept the distractions caused by our smartphones. Looking down at the screen and seeing it filled with fantastic beasts and realistic avatars is the next step in our acceptance of digital reality intruding into our analog lives.

# # #

BIG DATA AND PRIVACY

"Online privacy" is an oxymoron

April 28, 2017

At the Space Coast Tech Council (SCTC) quarterly dinner last week, the Big Data Florida team held a panel session on various aspects of big data. I provided an overview of big data, with an emphasis on data volume. Karl Seiler spoke about where big data comes from and role of the Internet-of-Things (IoT). Tauhida Parveen spoke about predictive analytics and how these techniques can help businesses identify trends in customer activity.

After the presentations, there was a lot of discussion about privacy and big data, perhaps reflecting one of the tag lines for the event: "What does big data know about you?" Turns out, the answer is, "A lot."

There is a constant tug of war between the positive benefits afforded by new technology and the tradeoffs that society is prepared to accept to realize those benefits. Privacy is a prime example of that conundrum. For example, if you enjoy using Facebook, you should be aware of how much personal information you are releasing online. Your digital footprint is significant, even if you're not fully aware of making the impression. And unlike real footprints, the online version never goes away.

For some people, privacy is very important. For others, the line between private and public vanished long ago. Scott McNealy, the past CEO of Sun Microsystems, was famously quoted as saying, "You have zero privacy anyway. Get over it." I believe this statement is becoming increasingly accurate. You don't even have to go online to become part of the big data collection – you just have to live.

Consider the upcoming UEFA Champions League soccer finals in Wales this June. As reported in The Verge, the police will be conducting a massive automated face recognition experiment that aims to scan every person in and around the Cardiff stadium, including the train station, city streets, and other public locations. They are expecting over 170,000 visitors, and each will be compared to a security database in real time. They are ostensibly looking for "people of interest," but it's inevitable that mistakes will be made. The software is good, but not that good.

It's an open question what will happen to all those images once the games are over. For example, will the database be shared with other agencies? Countries? Private companies?

The use of surveillance cameras by various agencies is not new. What is new is the scale of this operation and the speed at which it will operate.

If you thought getting physically hauled off a plane was bad, think about getting tackled by riot police just because some errant AI thought you looked a bit like the bad guys.

# # #

MACHINE INTELLIGENCE

Deep Blue, Watson, and AlphaGo

May 5, 2017

On May 11, 1997, machine beat man. The game was chess. The machine was IBM's Deep Blue. The man was world champion Garry Kasparov. It took just over an hour for Deep Blue to win the final game of the tournament.

Chess was long thought to be a game where humans would remain dominant. There are so many possible moves and complicated strategies used by grandmasters that a simple machine seemed a poor adversary. But Deep Blue was no simple machine.

IBM's supercomputer was a massively parallel system that could evaluate 200 million positions per second, searching six or more moves deep each time. Its programmers had given it a significant opening library, based on input from several chess experts. Kasparov was fighting not just Deep Blue the machine, but the accumulated knowledge of several grandmasters that Deep Blue had acquired.

But was Deep Blue intelligent?

Skip ahead nearly 15 years. In 2011, IBM's Watson supercomputer beat the reigning champion of Jeopardy!, Ken Jennings, to a much larger national audience. Jeopardy! is a much different game than chess, and in many ways it's more challenging for a machine. There are many nuances and cultural references in the questions, which means brute force analysis is not quite as effective. This is why the loss to a machine was seen as such a big deal: humans were supposed to be better at this.

But was Watson intelligent?

Skip ahead another 5 years. In March 2016, Google's DeepMind project demonstrated mastery of Go by beating a 9-dan professional in a five-game match with AlphaGo. The computing community had long believed that human Go champions would remain unbeatable by machines, due in part to the huge number of possible moves on the Go board, but rapid advancements in hardware and software proved the experts wrong.

But was AlphaGo intelligent?

The answer to all three questions depends on how we define intelligence. Deep Blue, Watson, and AlphaGo all achieved results that, at the time, were seen as revolutionary. They beat the leading humans in their fields. It might not have been a fair fight, but they won nevertheless. In that sense, these machines could be considered intelligent.

Some would argue that the machines were not intelligent – their human builders were. There is truth to that viewpoint. But AlphaGo relied on machine learning techniques that mimic how the human brain works. Seen from the outside, as a black box, AlphaGo functioned like a human brain – only better.

Which leads us back to Watson. IBM's current computer has advanced since it's 2011 TV appearance. In a sense, it has evolved. Have we?

# # #

WOMEN IN STEM – CHALLENGES

Cracking the gender code

May 12, 2017

I recently attended a seminar called "Women in Computing: Challenges and Opportunities" by the IEEE Computer Society. The seminar highlighted concerns specifically related to computing as a career choice for women – and the concerns are quite real. For example, the recent study "Cracking the Gender Code" reported that by 2025, the number of women in the computing workforce could decline to 22%; in 1995 that number was 37%. Not only would this be a terrible waste of human capital, it would have serious negative consequences for our economy.

More generally, women in many of the STEM (science, technology, engineering, mathematics) fields continue to struggle to gain footing. Even after numerous attempts to boost enrollment in schools, to provide positive role models for budding young scientists and engineers, and to encourage young girls to pursue careers in STEM, there are still societal problems. The recent book *Lean In* by the COO of Facebook, Sheryl Sandberg, illustrates some of these issues at the highest corporate levels.

One of the challenges is getting more women interested in STEM in the first place. This means encouraging young women when they are still in the K-12 environment. If we wait until they enroll in university, it's probably too late. I personally think a focus on coding alone is not the best way of stimulating interest in a computing career. Computing is about far more than just programming; it's about problem solving, something that can be demonstrated across multiple disciplines.

A second challenge is keeping women interested in STEM once they enter the workforce. It's undeniable that women have different challenges than men in male-dominated areas like computing. The often-cited work/life balance is something that most people in STEM struggle with, in part because they are passionate about their job. In general, our society still implicitly expects women to do more for the family, which inevitably leads to a shortage of time. Not everyone wants to pull all-nighters at work, but the unfortunate image of many STEM careers is of hackers spending all their time in front of a screen, or of a scientist never leaving the lab.

A third challenge is nurturing women who choose to remain in STEM fields to become leaders and mentors. Learning leadership skills is an essential component of career advancement into management roles. Fortunately, these are skills that can be learned over time. But for women, there are clearly additional stumbling blocks. For example, many people view vocal women as aggressive, where they would view the same behavior from a man as confidence. The military has dealt with this issue for years; the technology world needs to catch up.

# # #

Women in STEM – Opportunities

Eliminate negative stereotypes

May 19, 2017

Last week I wrote about three challenges for women in STEM (science, technology, engineering, mathematics) fields. The challenges were getting more women to choose a career in STEM, keeping them interested in STEM once they enter the workforce, and nurturing the next generation of female STEM leaders and mentors. Each of these challenges can be viewed as opportunities to increase the number of women in STEM.

Consider the first challenge: encouraging more women to choose a career in STEM. This needs to be done at the K-12 level, while young people are still flexible in selecting their career path. There are many advantages that working in STEM affords, such as earning high salaries, having the opportunity to work with other smart people on really interesting projects, and enjoying the feeling of having a rewarding career and not just a mundane job.

But there is one major issue that prevents young women from realizing these advantages: how people in STEM – and in particular, men in computing – are portrayed in the media. Why is it that on many TV shows, lawyers wear suits, doctors wear white coats, but "the IT guy" (and it's almost always a guy) is shown as socially awkward, dorky, and working in a windowless basement? Just think of the character played by Jorge Garcia in the "Hawaii Five-O" reboot. Or any of the men on "Big Bang Theory." Who would want to be geeky outcasts like them? Sadly, there are many such examples that highlight negative stereotypes of what working in STEM is like, and unsurprisingly, impressionable young girls are turned off.

For the second challenge of keeping women interested in STEM careers once they enter the workforce, the opportunity is to correct systemic issues in many companies' corporate culture. A few months ago, a blog post by Susan Fowler about her time as a female engineer at Uber went viral. The discrimination and misogyny she experienced was alarmingly blatant for such a modern high-tech company. But from the many subsequent posts by female employees at other companies, it seems her story was not unique.

For the third challenge of nurturing the next generation of female STEM leaders and mentors, there are several good places to start. For example, in the computing field, attending the Grace Hopper conference is a wonderful opportunity for women to network with other women who are working in the same field, but at different stages of their careers, from graduate students to senior faculty. There are presentations on work/life balance, professional development, and leadership skills. But most of all, the conference is an opportunity to hear success stories and return home inspired.

# # #

WOMEN IN STEM – INTERVIEW WITH TAUHIDA PARVEEN

Keys to success for women in STEM

May 26, 2017

The past two columns have dealt with some of the challenges and opportunities for women in STEM (science, technology, engineering, and mathematics) careers. Tonight, the Center for Technology & Society presents the final "Tech Talk" of the 2016-2017 season at 8:00pm in the Henegar Center. The topic is "How Women Can Succeed in STEM." The guest speaker is Dr. Tauhida Parveen, who will relate her experiences working in technical and leadership positions in academia and industry.

Dr. Parveen is University Department Chair of Software Engineering at Keiser University. She is also Lead Instructor at Thinkful, a NYC-based startup focused on the online education experience for tomorrow's developers. She holds a PhD in Computer Science from the Florida Institute of Technology and an MBA from the University of Central Florida.

In preparation for tonight's presentation, I sat down and spoke with Dr. Parveen about some of the key points she'll cover in her talk.

Technology Today: What do you think are important personal characteristics women should develop to succeed in STEM?

Tauhida Parveen: I think it's very important for women to learn to be self-reliant. Just be confident that you can succeed. You don't need to be an 'A' student either; persistence is more important.

I also think women should stop thinking of themselves as women first. Instead, think of yourself as a person first. I rarely enter a room full of engineers worrying that I'm one of the few females present; I just don't think that way.

Lastly, don't be afraid to fail. You'll never get anywhere without trying new things, and worrying about what others may think of you is a waste of time. When you see success in others, you don't see all the failures they also experienced.

Technology Today: As a woman of color working in a STEM field, have you experienced discrimination in the workplace, and if so, how have you dealt with it?

Tauhida Parveen: Yes, occasionally. There is still an "old boys" network in many STEM fields, including computing. I generally ignore it. You can't sweat the small stuff in your life. I can't control what other people say or do – I can only control my reaction to them. Thankfully, most of my younger colleagues don't act like that; they value abilities and contributions to the team, not gender or background.

Technology Today: What advice do you have for young girls thinking about a career in STEM?

Tauhida Parveen: Go for it! My choice of STEM for a career was one of the best decisions of my life. You have a chance to make a difference in the world. What more could you want?

###

SUMMER READING

Time to wind down with a few good books

June 2, 2017

Memorial Day has come and gone, which means we're officially heading into summer, and with the start of summer comes my annual summer reading list. This year, I've selected five very different books that share a theme of the role of technology in our lives. Any one of the books would be an excellent read at the beach.

The first book is *Bowling Alone: The Collapse and Revival of American Community* by Robert Putnam (Simon & Schuster, 2000). This is not a new book, but the exhaustive analysis by the author on the societal changes over the last few decades brought on by devices such as television are prescient. I've personally experienced the difficulty in getting people to participate in community organizations, and it's getting even more challenging with the continued use of social media as a replacement for actual social interaction.

The second book is *The Dumbest Generation: How the Digital Age Stupefies Young Americans and Jeopardizes Our Future* by Mark Bauerlein (Tarcher/Penguin, 2008). This is a fascinating but rather depressing book about the mindset of Millennials and how their habits will affect society in the near future. Just ask any teacher.

The third book is *The Death of Expertise: The Campaign Against Established Knowledge and Why It Matters* by Tom Nichols (Oxford University Press, 2017). I couldn't put this book down. The author writes about technological advances and the negative impacts they are having on society, in particular "a surge in narcissistic and misguided intellectual egalitarianism." The use of "alternate facts" in the current political arena is a good example of this.

The fourth book is *Our Final Invention: Artificial Intelligence and the End of the Human Era* by James Barrat (Thomas Dunne Books, 2013). The author does an excellent job discussing topics that I've spoken about a lot in the past year, such as the rise of intelligent robots and how this will dramatically change our society. Scary stuff.

The fifth book is *Chapter and Verse: New Order, Joy Division, and Me* by Bernard Sumner (Thomas Dunne Books, 2014). This autobiography from the lead singer of the band New Order offers some fascinating insights into how the group changed their musical style over time, partly in response to new digital instruments. The author explains the role of music technology in creating hits like "Blue Monday" in the 1980s.

There is one last book I'm looking forward to reading: *Win Bigly: Persuasion in a World Where Facts Don't Matter* by Scott Adams (Portfolio, 2017). This timely book from the creator of Dilbert is scheduled for release this October. You can read early excerpts on Mr. Adams blog.

# # #

APPLE WWDC 2017

Machine learning was everywhere

June 9, 2017

This week, Apple held its Worldwide Developer Conference (WWDC) in San Jose, Calif. CEO Tim Cook led the audience through a tour of six new products or enhancements. There were updates to their entire operating system line across multiple devices (computer, phone, TV, watch), upgrades to their Mac computers, and enhancements to several common applications (e.g., Safari, Photo). The only thing truly new was a home speaker system, rather like Amazon.com's Echo, that won't ship until December. It was a rather pedestrian event that seemed to lack excitement.

One thing that really struck me was how many times the Apple presenters said the words "machine learning." For example, Apple's Senior Vice President of Software Engineering, Craig Federighi, mentioned machine learning in the context of new hardware support for faster processing, how it's used for facial recognition in Photo, and how machine learning underpins all the advancement in Siri. Apple has been playing catch-up in the artificial intelligence world, but it appears they're doubling-down on the resources needed to compete against companies like Google and Amazon.com.

Virtual reality was also mentioned several times during the keynote presentation. There are many rumors that the next iPhone will support augmented reality, but it was virtual reality that was highlighted several times. This is probably more important for developers than for users, as much of the emphasis was on gaming engines.

Another announcement of interest was Amazon Prime Video coming to the Apple TV. Just as there are apps on the Apple TV for services such as Hulu and Netflix, there should soon be an app for Prime Video. This is a big deal, because now you need to have a separate device (Amazon Fire TV or Stick) to play Amazon Prime Video content on your TV. By making Prime Video accessible through an app, the additional hardware will no longer be needed. I'm not entirely sure why Amazon is doing this, but I assume they make more money from their content than they do selling hardware. I'm also surprised that Apple is doing it, because Prime Video is a direct competitor to iTunes on the Apple TV, but perhaps the repeated requests from their users finally paid off.

In addition to Apple's WWDC, there are other developer conferences held each year, including Microsoft Build, Google I/O, and the Facebook Developer Conference. If you have a chance, go to one of these events next year. It will give you an inside peek into the mindset of the developers who create the computing tools and toys we've come to rely upon. It's not always pretty, but it is fascinating.

#

FATHER'S DAY

A tablet, a browser, and YouTube

June 16, 2017

If you're still wondering what to get dear old dad (or granddad) for Father's Day this year, consider giving him the ability to take a digital walk down memory lane. If he already has a computer and is comfortable using it, then this is probably not for him. But if he's like one of the many older folks who still doesn't use computers very much, then this is a relatively painless way of getting him online and doing more of what he already likes to do: watch TV.

One of the reasons people avoid using computers is that they are still far too difficult to use. But a stripped-down tablet can be tailored for use by just about anyone. The tablet could be the latest iPad Pro, but it could just as easily be a far less-expensive Kindle Fire. It only needs two apps installed: a browser and YouTube, both of which are available across all devices and platforms. These two apps are also available on most Smart TVs, but search is more cumbersome, which is why I recommend a personalized tablet instead.

YouTube offers a wealth of channels and programs. The quality varies significantly, but the ability to watch old TV shows is something dad will appreciate. For example, a quick search on YouTube for "I Love Lucy" brings up many episodes of this timeless classic. The same is true for many of the shows dad used to watch, which he can now enjoy again.

The browser is used primarily to access streaming media services such as Netflix and Amazon Prime Video. You'll have to setup an account for dad on each of these services, but once that's done, he'll be able to watch movies and TV shows at his leisure. He'll also be able to listen to audio books, Internet radio stations, and live TV if he wants to. Browsers like Chrome and Safari all support this, so the tablet he'll use is not very important. The Fire 7 tablet from Amazon is just $50 and it comes with Alexa as well, so dad can ask questions about the weather or sports scores in a natural manner.

I suggest you remove (or bundle into a "Do Not Use" folder) all the other apps that come on the tablet. Keep the interface clean, with just the browser and YouTube on the main screen. Once dad gets used to it, you can expand the tablet's capabilities to include photos, something that is sure to please him. It's all about making it easy for him to relive the best things of his past from the comfort of his favorite chair.

# # #

TOO HOT TO FLY

Technology can't beat the heat

June 23, 2017

PHOENIX, Ariz. – Snow. Ice. Hurricanes. These are the sorts of natural events that can affect travel. Unfortunately, I've experienced all of them. Now I can add another type of weather-related travel interruption: heat.

I arrived in Phoenix on Monday afternoon when it was already extremely hot. American Airlines was warning passengers to call ahead to verify that their flights would be departing on time. On Tuesday, AA canceled several dozen flights due to the excessive heat at Sky Harbor airport. It was a record high of 119 degrees yesterday. Today is much the same and looks likely to break another record.

The jets affected by the heat are smaller regional aircraft, such as the Bombardier CRJ, which have a maximum operating temperature of 118 degrees. Even if the planes could function, they'd need a much longer runway for liftoff in the dry air, and PHX doesn't have runways long enough for these aircraft. This is not a situation that most people ever think will happen – but it does. Airport maintenance personnel can shovel the runways clear of snow and they can de-ice the aircraft wings, but they can't change the laws of physics when it comes to heat, speed, and lift. It was just too hot to fly.

I wonder if the CRJ has this limit because Bombardier is based in Montreal, and Canadians are more experienced with cold wintery weather, not the oven-like temperatures here in the Valley of the Sun.

It's a healthy reminder that technology isn't infallible, and relying on gadgets in extreme conditions can be a recipe for disaster. For example, my insulin pump is near the edge of its temperature limits, and the insulin itself will break down if I stay outside too long. All electronics have an operational envelope and this week that envelope was torn apart. If the global climate really is warming up, then lessons learned from places like Phoenix will be invaluable as temperatures rise elsewhere in the world.

The highest recorded temperature ever in Phoenix was 122 degrees on June 26, 1990. This is getting perilously close to the maximum operating temperature of the larger jets from Airbus and Boeing, which is about 126 degrees. That sounds unbelievably hot, and it is, but it actually gets hotter in the Southwest. Today's forecast for Death Valley, California is 128 degrees – and that's in the shade. In the direct sun, it's hot enough to force your iPhone to warn you that it needs to cool down before it can be used. You'd be barbequed before you can take a selfie, which seems like a suitable beginning to a long, hot summer.

# # #

Canada 150

Take off, eh!

June 30, 2017

Canada's 150 birthday is July 1. To help celebrate, I recommend eating some poutine, washing it town with a cold Ex, and topping it off with a beaver tail. If those delicacies sound foreign to you, no worries – we can also celebrate the technology contributions from my home, the Great White North!

Everyone knows Canada is the source of winter cold fronts, but it's also the birthplace of ice hockey, Dudley Do-Right and the Mounties, and extreme politeness. It's also given us Captain Kirk (William Shatner), Austin Powers (Mike Myers), and countless Titanic karaoke performances (Celine Dion – sorry about that last one).

Canada is not just home base for countless comedians, musicians, and athletes now in the United States. It's also where some of our most important scientific inventions came from. Let's start with insulin, a discovery that literally saved my life. The 1923 Nobel prize in medicine was awarded to Banting (working with Best) and Macleod in Toronto for their research on injectable insulin. Until their work, Type I (insulin-dependent) diabetes was a death sentence. We now use synthetic insulin that closely mimics the body's natural insulin, created through a process of recombinant DNA manufacturing, to help manage this chronic disease.

Skipping ahead over half a century, and switching domains from medicine to aerospace, Canadian engineers developed the Shuttle Remote Manipulator System (SRMS), which is better known as the

Canadarm. It is a collection of robotic arms that were used on the Space Shuttle orbiters to manipulate large payloads. The newer version, Canadarm2, is currently used on the International Space Station. Robotics is increasingly important for automated space travel and the Canadarm has led the way.

If you enjoy large-screen summer movies in IMAX format, you can thank the pioneering work from the company's founders at Expo 67. The Expo took place in Montreal to celebrate Canada's centenary. I still vaguely recall being pushed in a stroller around the busy complex (I have clearer memories of Expo 86 in Vancouver), particularly the Biosphere, a giant metallic globe that housed the United States' pavilion for the 1967 world fair. Today, IMAX theaters are commonplace.

Lastly, we can thank Alexander Graham Bell for inventing the telephone in 1876. His famous words to his assistant were the first ever to be transmitted electronically: "Mr. Watson, come here, I want to see you." Little did Bell know that over a hundred years later his new device would power a communications revolution and an app ecosystem.

This weekend I'm going to binge watch some "Rocky and Bullwinkle." Maybe have a smoked meat sandwich and a Nanaimo bar for dessert. Good idea, eh?

\# \# \#

GAME OF THRONES

Winter is coming

July 7, 2017

Winter is coming. It may seem strange to say that, considering summer has barely begun and we have several months of blistering heat and muggy weather still to endure, but technically it's true. After the summer solstice, the days start growing shorter and by December the temperatures here in Florida will have dropped into the chilly 80s.

But in Westeros, the phrase "winter is coming" has a decidedly different connotation. For those of you who are not fans of HBO's incredibly successful series "Game of Thrones," Westeros is the continent where most of the complicated story arcs take place. The series is based on the first several books by author George R.R. Martin. It returns to the screen next week with its penultimate season.

Westeros is ruled by the Seven Kingdoms, led by whomever sits on the Iron Throne in King's Landing. The kingdoms are rife with internal fighting, each faction constantly scheming to gain more power. It's rather like a medieval version of "House of Cards."

Westeros faces an existential threat from White Walkers, an ancient race with magical powers that have the ability to resurrect the dead from the battlefield and force them into undead servitude. White Walkers live in the frozen north, beyond the Wall, a massive barrier manned by guards of the Night's Watch.

Society in Westeros has not changed in thousands of years. There is a ruling class in each separate kingdom, various religious orders and civil servants, with the rest of the population basically serfs or slaves. Because of this structure, there is little incentive to innovate. The result is that Westeros civilization has not really advanced for millennia.

Contrast this stagnation with our society. Since the industrial revolution, we've made incredible technological advancements. In just a few hundred years we've moved from an agrarian lifestyle of a modern, connected world. We may worry about climate change, but in Westeros, winter doesn't have a fixed length; sometimes it lasts for decades – which makes growing crops challenging.

When the White Walkers inevitably breach the Wall, what will the Seven Kingdoms do? What technology do they have to repel the invaders? Cross bows, some unstable green Napalm-like substance, and maybe a few jumbo-sized dragons that are difficult to control. Good luck with that.

We have our own external threats (e.g., North Korea), but thankfully we have more options to deal with them. To paraphrase John Lennon, imagine there was no technology. We'd be stuck throwing rocks and hoping for the best. So, the next time you find technological change too fast to keep up with, remember the alternative. Winter is always coming.

###

PROGRAMS VS. PROGRAMMERS

Who should be trained as the next generation of developers?

July 14, 2017

JACKSON HOLE, Wyoming – This is cowboy country. There are elk all over the place. Bison and wolves roam the nearby parks. The Grand Teton mountain range dominates the scenery. People talk about horses and fishing constantly. What they don't talk about is programming, and that's too bad. Cowboy poetry is quite popular; cowboy software could be too.

For several years there have been warning signs that universities are not producing enough programmers to fill all the open jobs that power our modern economy. Tens of thousands of positions go unfilled while humanities graduates become baristas. Even the proliferation of online certificate programs has failed to make much of a dent.

There are several laudable initiatives that try to re-train people for new careers, such as Coalfield Development Corp. in West Virginia that focuses on Appalachians who lost their traditional jobs and are now preparing for new careers in fields such as solar panel installation or woodworking. These are innovative attempts to give meaning to people's lives and to boost the employment rate in historically challenged areas. But they're not enough. We need something different, a new approach to solving the programmer shortage.

Maybe the solution is to stop training people to be programmers and start training programs to be programmers.

"No way," I can hear people say. "Programming requires extensive technical knowledge and a flair for creativity that just can't be automated. It can't be done."

"Way," I say. It's already being done.

We've had automated programming systems for decades. When I was developing compilers at IBM we used program generators that created scanners and parsers based on detailed language models. These generators are called "compiler compilers." They are programs that produce other programs. They have been used in areas that are well-understood and the solutions can be codified for easy replication. They are not creative, but they are efficient.

Now, advances in artificial intelligence have created a growing market for intelligent programs that can solve problems previously thought to be unsolvable by anyone but a trained software engineer or computer scientist. For example, the company DiffBlue markets tools that do automatic test case generation, code refactoring, and deep security analysis of large applications. DiffBlue is a spinoff from Oxford University, led by professor Daniel Kroening. They recently received $22 million in Series A capital funding, which means the market clearly believes their products have value. DiffBlue relies on machine learning techniques to perform some pretty impressive analyses.

Of course, there's still employment opportunities to be had, such as working for DiffBlue to improve their toolset. For now, anyway, before the tools learn how to improve themselves.

#

OK Computer

Hey man, slow down

July 21, 2017

The latest installment of the rebooted "Planet of the Apes" franchise just opened. It's yet another storyline depicting a dystopian future for mankind. Clearly, we like reveling in misery, because these movies continue to be popular.

Twenty years ago this summer, the band Radiohead released the album "OK Computer." Many critics still hail it as the definitive soundtrack to the 1990s: bleak, foreboding, and technically experimental. The band's lead singer, Thom Yorke, has a uniquely ethereal falsetto voice that sounds like distant wailings emanating from something locked deep inside a machine.

The album was recorded in a thousand-year-old manor in Bath, England. Yorke claims it was haunted, which perhaps explains some of the lyrics. He has said "OK Computer" was primarily inspired by near-constant touring and a feeling of alienation from the regular world. But music, like poetry, is open to interpretation by the consumer, and listeners have found a deeper message about isolation through technology in the record. Songs like "Paranoid Android" (named after Marvin, the depressed robot from "The Hitchhikers Guide to the Galaxy") really resonated with their fans – and in some respects the message is even more germane today.

"OK Computer" is a concept album of sorts, and one of the last before digital downloads and streaming services fundamentally disrupted the music distribution model. Remember that 1997 was the time of AOL and Windows 95. The Internet was in its infancy. Social media didn't exist. But even then, Yorke felt a sense of dislocation and loneliness, and he worried that rapidly advancing technology would only make it worse. How prophetic. Today we're ever more connected and ever more alone.

The album begins with "Airbag," a commentary on our over-reliance and misplaced trust in technology. I wonder what Yorke would have written if self-driving cars and robotic automation were as mainstream then as they are now.

The song "Karma Police" depicts a surveillance society in the UK – which now bristles with security cameras at virtually every intersection in the country. In the song "Fitter Happier," a disembodied computer voice channels Kafka to describe humans as bugs in a cage. Is technology forcing us to undergo our own metamorphosis? The song "Electioneering" foreshadows government coercion – and now we're awash in fake news and predictive analytics that model people as malleable data sets.

The album closes with "The Tourist." The song has a line that couldn't be more relevant to our harried lives today: "Hey man, slow down." But I can't slow down: I have alarms, notifications, calendar reminders, popups, push alerts, Twitter feeds, and Facebook updates constantly in my face 24/7. Who has time to slow down?

# # #

LONGEVITY

Rocking into the 80s

July 28, 2017

Medical technology has made incredible advancements in the last few decades. There are modern pharmaceuticals to alleviate the symptoms of almost anything (even though they sometimes cause more problems through misuse, such as opioids). Joint replacements give people a new lease on mobility and independence. We may soon have personalized gene therapy to counter the effects of aging itself.

The result of these advancements is that we are living longer than ever, which is changing our society in fundamental ways. A recent article in *The Economist* says a new age category is needed to describe people over 65. "Geriactives"? "Sunsetters"? "Owls" (Older, Working Less, Still earning)?

According to the article, "Before 1800, no country in the world had a life expectancy beyond 40. Today there is not a country that does not. Since 1900, more years have been added to human life than the rest of history combined, initially by reducing child mortality and lately by stretching lifespans. Longevity is one of humanity's great accomplishments."

To illustrate the "new old," the article features a photo of the Rolling Stones in concert last October. Their lead singer, Sir Mick Jagger, is 73 years old. He looks a bit leathery, but no one could say he lacks energy on stage. And he still seems able to manage the fatigue that comes from the demands of constant travel. He may not be a typical senior, but he may be the new normal.

65 has traditionally been the age we think of as retirement. Some researchers say 115 is the maximum age the human body can reasonably expect to function. Other researchers say there is no known limit, and rapid advances in gene therapy may make old age a thing of the past. If we're going to live for another 50 years, we clearly need to rethink this notion.

If everyone will live to be over 100 years old, how does that change the notion of marriage? There has been a surge in "silver divorces," where people choose to go their separate ways in their 60s (or later), even after decades together. They may feel that if there are many good years left, the risks of being alone are diminished.

Think about employment opportunities and career paths over your extended lifetime. How could your four years at college when you were 20 prepare you for life 90 years later? The notion of life-long learning will become commonplace, which may disrupt the traditional education models we have now.

If people didn't work well past 65, who would fund our social programs?

And besides, what would you do with all that free time?

# # #

GENOME EDITING

Return to the island of Doctor Moreau

August 4, 2017

H.G. Wells published his novel "The Island of Doctor Moreau" in 1896. The plot focuses on a shipwrecked English sailor who is rescued and taken to a remote island in the South Pacific. The island's inhabitants are animal/human hybrids. They are created through painful surgical procedures by Doctor Moreau, who is trying to perfect the process of transforming animals into humans. The hybrids inevitably revert to animal form. The sailor escapes the island, but not before the hybrids return to the jungle and the Doctor's compound burns to the ground.

I was reminded of Wells' incredible imagination this week when I heard that scientists had successfully edited the DNA of human embryos to remove a potentially fatal heart condition. This means that, for the first time, humans are able to edit (add, delete, or change) their own genome.

Technically, this development is an incredible scientific advancement. The tool used for genome editing is called CRISPR-Cas9. It has proven to be a very efficient mechanism to perform complex genetic manipulations.

The potential medical benefits from genome editing are significant. We might be able to edit a fetus' DNA to remove inherited defects before birth. We might also be able to craft personalized medical procedures that are tailored to each person's DNA, which could help alleviate dangerous side effects from pharmaceuticals while increasing the drugs' efficacy. As I wrote in this column last week, we might even be able to increase longevity by slowing down or even halting the aging process through gene therapy.

The ethical implications of genome editing are equally significant. Performing experiments on human embryos is already the subject of much debate. These new procedures have the potential to more dramatically alter living tissue, which could lead to unintended moral dilemmas.

If genome editing becomes commonplace, couples could procure designer babies, with all the physical traits they desire. Blue eyes? Lighter skin tone? Increased muscle mass? No problem. But what would stop some rogue clinic from becoming the modern-day Island of Doctor Moreau?

Are you feeling a little more lethargic as you get older? Let's just tweak your genome a bit by inserting a slice of leopard DNA. Done! Want better eyesight? A bit of eagle eye blended into your genome and you're good to go. This may sound farfetched, but genome editing is just in its infancy.

Most pundits assumed humans would be augmented with mechanical aids to improve performance, including brain capabilities. In the genome editing model, we're changing biologically instead. Like rapid advances in artificial intelligence, researchers may have opened a Pandora's Box that could lead us to places we can't begin to understand.

# # #

BACK TO SCHOOL

Technology to help smooth the experience

August 11, 2017

It still feels like we're in the middle of summer, but here in Florida, it's time to go back to school. Everyone from elementary school students to fresh university undergraduates are getting ready to kick off a new year of study. When I was young, I looked forward to getting back into the classroom, but I know not everyone shares that feeling of enthusiasm. For many students, August is a time to dread – but it need not be that way. To make things feel better, why not try a little retail therapy? (Sorry parents!)

Let's assume you are entering college, living in a dorm on campus. What technology can you buy that will improve your educational and social experience without breaking the bank? I can suggest three things: headphones, a tablet, and a vacuum.

Living with other people in the same apartment for the first time can be an unnerving experience. Living with fellow students means sharing the kitchen, the living room, and putting up with their eccentric habits – like playing loud music at 3:00 am. A nice pair of headphones will let you tune out the world. It won't matter what noises are coming from the other rooms: you'll be safely ensconced in soothing music (or audio books). Furthermore, the headphones will make you popular with your roommates as well since they won't be subjected to your own musical selections.

Tablet computers have become very inexpensive. They are a great way to consume streaming media, such as watching movies wherever there's a Wi-Fi signal. That means you don't need to fight over the communal TV. If you want to watch "Game of Thrones," and your roommates are set on watching "Dancing with the Stars," you can catch up with all the news from Westeros without any arguments. Of course, tablets can also be used to do real work, such as writing reports and doing online research. If you get a stylus, they're perfect for taking notes in class too.

Most student accommodations start to look like rubbish tips by the time midterm exams roll around. It's not that everyone tries to mimic the movie "Animal House." It's just that student schedules can get very busy. Who wants to waste valuable time cleaning up after all those parties? Fortunately, we have robot vacuums now that work tirelessly to keep your floors clean. They can be pricey, so it might be a good item to purchase jointly with your roommates.

If you're entrepreneurial, maybe you can rent out your robot to friends for a little cash on the side. It's cheaper than maid service and more fun to watch.

# # #

Voyager at 40

The <u>real</u> Interstellar

August 18, 2017

Voyager 2 launched on August 20, 1977, from Cape Canaveral, with Voyager 1 closely following it on September 5. I was not in Florida to see the launch in person, but I'm sure some readers will remember the blastoff of the Titan IIIE rockets. It was the start of an incredibly successful mission that continues today, nearly 40 years later.

Both Voyager spacecraft are still active, traveling through space over 10 billion miles from home. This is such an incredible distance that's hard for us to fathom. It takes radio signals from Voyager 1, traveling at the speed of light, nearly 17 hours to reach Earth. The signals from Voyager 1 are so weak that it's a monumental feat of engineering that we can even discern the signal. According to NASA, "The antennas must capture Voyager information from a signal so weak that the power striking the antenna is only 10^{-16} watts (1 part in 10 quadrillion). A modern-day electronic digital watch operates at a power level 20 billion times greater than this feeble level."

The two Voyager probes took different paths through our solar system, but they both have made important contributions to our knowledge of the outer planets and their moons. Some of their instruments, such as the camera, have been shut down, but others continue to operate. This will continue until about 2025, when Voyager 1's power supply will run down. The spacecraft will continue its journey, but we won't be a direct part of it anymore.

The movie "Interstellar," which was about astronauts traveling through a wormhole to reach interstellar space, came out in 2014. But in Voyager's case, science fact beat science fiction: Voyager 1 became the first manmade object to enter interstellar space in 2013. This was after passing through the outer boundary of our solar system, leaving the heliosphere and heading into unchartered territory. NASA named the Voyager project quite accurately: "The Interstellar Mission."

Both Voyager probes contain a golden record with images and audio from Earth. Famed astronomer Carl Sagan's voice is included, as is a message from President Carter. So far, no one has replied to our greetings, but on a cosmic scale, we've barely left home.

# # #

VOYAGER'S TECHNOLOGY

It's amazing what can be done with 68 KB of memory

August 25, 2017

When the Voyager probes were launched in 1977, the technology landscape was vastly different from what it is today. I'm typing this document on an Apple MacBook Air that weighs just a few pounds, yet it packs a powerful computing punch: dual 1.7 GHz processors, 8 GB of internal memory, and a 500 GB solid-state internal hard drive. I routinely enjoy over 100 Mbps wireless download (receive) speeds, and over 10 Mbps upload (transmit) speeds. And this is on a computer that's over three years old – a whole generation in computing terms.

The Voyager probes have to make do with far less computing power: less processing, less storage, and less networking. And yet they've been able to operate for over 40 years in the harsh environment of open space, exposed to extreme radiation during their flyby of Jupiter, and the fluctuating magnetic fields at the edge of the solar system. They are truly incredible feats of engineering.

There are three different types of computers on Voyager (each probe is identical). Each computer had two versions running, usually to support redundancy, but they were sometimes used together to increase computing power. As the mission aged, this redundancy was eliminated.

The first computer is the Computer Command System (CCS), which controls the overall spacecraft. It is similar to the earlier Viking system. The second is the Flight Data System (FDS). The third is the

Attitude and Articulation Control System (AACS). Each system was custom-built using components designed primarily at NASA JPL. The computers were first programmed in an early version of FORTRAN and assembly language.

Programmers today rarely think about how much memory their programs need. Even smaller applications, like those on a smartphone, basically assume infinite storage and rely on the operating system to manage things for them. The Voyager probes do not have that luxury: they must make do with just 68 KB of storage. That's about 117,000 times less memory than my MacBook.

The Voyager probes do not have any hard drive either. They rely on digital 8-track tape machines – rather like cars did for music in the late 1970s. The tape is reused when new data arrives.

When Voyager 1 transmits data back to Earth, it takes over 19 hours for the signal to arrive, and when it does, it's at very slow 160 bits per second. That's far, far less than the old dial-up modems we used to use. But then again, the signal is coming from over 12 billion miles away, sent from a 22-watt transmitter. Local radio stations broadcast at 50,000 watts.

It's amazing we can still communicate with the Voyagers at all.

# # #

VOYAGER'S GRAND TOUR

A cosmic alignment

September 1, 2017

Once every 175 years, a cosmic alignment occurs. The four outermost planets, Jupiter, Saturn, Uranus, and Neptune, become accessible to spacecraft from Earth. The massive gravitational forces of the planets are used to "slingshot" the spacecraft forward. This means there is no need for large amounts of onboard propellant or particularly powerful engines. Nature does the work for us.

This planetary alignment last happened in 1977, the year the Voyager probes were launched. NASA scientists started planning for this singular event in the 1960s. If anything went wrong, we would have missed our once-in-several-lifetimes chance of visiting the outer reaches of the solar system with two probes simultaneously.

Fortunately, everything went right, but there were a few glitches. According to the first "Voyager Mission Status Bulletin" published by NASA on August 9, 1977, the Attitude and Articulation Control System (AACS) and the Flight Data System (FDS) both experienced failures during testing of the VGR77-2 spacecraft – the one scheduled for launch on August 20. NASA switched to the VGR77-3 spacecraft instead, which became known as Voyager 2. This was only possible because there were actually three spacecraft assembled, but only two would fly.

Voyager 1 arrived at Jupiter on March 5, 1979, followed a few months later by Voyager 2 on July 9, 1979. After exploring the Jovian system, both probes continued on to Saturn, with Voyager 1 arriving at the ringed planet on November 12, 1980, and Voyager 2 arriving the next year, on August 25, 1981.

After Saturn, Voyager 1's trajectory took it away from the outer planets and into interstellar space. Voyager 2 continued its path towards Uranus, where it arrived on January 24, 1986. It really puts distance into perspective when you realize that it only took Voyager 2 about two years to reach Jupiter, but it took another seven years to reach Uranus. The outer planets really are "out there."

Voyager 2's final planetary encounter was with Neptune on August 25, 1989. The probe had traveled about 2.8 billion miles in 12 years, but that was really just the beginning of its journey. Like Voyager 1 before it, Voyager 2 headed towards interstellar space. It's now over 10 billion miles from home, somewhere on the boundary of the solar system. Voyager 1 is over 12 billion miles away and has already crossed the heliopause into uncharted territory.

When we are long gone, the Voyager probes will continue their grand tour of outer space without us. Some part of humanity may someday reach the stars. Who knows, maybe the storyline from the first "Star Trek" movie will come to pass, and Voyager will return home.

# # #

HURRICANE IRMA

Technology to help you stay calm during the storm

September 8, 2017

Large storms like Hurricane Irma can make you nervous. You're never quite sure where or when the hurricane will make landfall. Should you stay and ride it out, or evacuate to a shelter? Will your home sustain any damage? Will you lose power?

The last worry, losing power, is increasingly a major concern. Without power, most of our technology wouldn't work. Never mind that the air conditioning would fail and you'd be sweltering in the heat and humidity – you'd not be able to post Facebook updates or send out tweets. Unacceptable!

I've been in Florida when a hurricane strikes and knocks out power for several days. It's extremely stressful and physically uncomfortable. By Day 3, people are reverting to a "Lord of the Flies" existence. It reveals our very thin social veneer.

But technology can also help you stay calm during the storm. Think about the likely sequence of events that will occur when the hurricane arrives, and plan accordingly. After you've safeguarded your house, your belongings, and your family as best you can, relax with a movie. Try to ignore the howling winds outside.

If the cable or Internet goes off, pop in a DVD and watch something offline. You don't really need the news at this point anyway; the hurricane is already here – what are you going to do about it, other than try to relax?

If you don't have any DVDs, try watching TV using an antenna. Most stations will continue to broadcast over the air. The only downside to this mode of entertainment is the advertisements you're forced to sit through. And the incessant hurricane news flashes.

If movies aren't your thing, but you like music, prepare your iPod with a selection of soothing tunes. This is a time when a streaming service is not as good as having an actual copy of the music on a local device.

If you prefer to read, download a few books and magazines to your iPad or your Kindle before the storm arrives. Even if the power fails, you can still read your material until the battery dies. At that point, go old-school and read an actual book.

If power is not available in the house, make sure you have a car adapter to charge your portable devices once it's safe to go outside.

I find that constantly checking websites and weather apps for storm updates just makes things worse. By all means, stay up-to-date, but don't become addicted to hurricane news every five minutes – it only agitates you. Remember, you can't control the storm; you can only control how you react to it.

Stay safe!

#

FACE ID

Who owns your facial recognition data?

September 15, 2017

According to the September 9-15 issue of *The Economist* magazine, "Facial recognition is not just another technology. It will change society."

During the Apple Special Event on September 12, CEO Tim Cook said, "Technology infused with humanity could improve people's lives and change the world. No other device in our lifetimes has had the impact on the world that the iPhone has."

With the upcoming release of the iPhone X, Apple promises to marry a technology that will change society with a smartphone platform that has already impacted society more than anything else in recent history. This could indeed be a game changer.

The capabilities offered by Face ID, Apple's facial recognition hardware/software solution in the new iPhone, will let you open your phone just by looking at it. No more PIN codes. No more swiping. No more fingerprints with Touch ID. Just a quick glance at the screen and you're good to go. The convenience seems unmistakable.

We can skip the technical details concerning the workings of Face ID. Suffice it to say, the phone scans your face using multiple cameras, and compares the image to models it has created. The comparison is powered by a new A11 Bionic processor that contains dedicated neural networks to rapidly analyze thousands of data points. The comparison is supposed to be impervious to changes in your appearance, such as wearing glasses, growing a beard, or changing hairstyle. It's even supposed to be able to detect if you're trying to fool it with a photo or a mask.

Apple says the data is on the phone, never on the cloud. But everything can be hacked with enough effort and ingenuity. Consider the recent massive data breach at Equifax, where hackers gained access to personal data from 143 million citizens – all due to a known security flaw that was not patched on a website.

This leads to an important question: who owns your facial data? What rights do you have to control its use? Could it be accessed by third party apps without your explicit consent?

Law enforcement and national security agencies would love every citizen to voluntarily provide facial data, just like finger prints. They are currently limited to searching visual databases of known offenders, federal employees, and so on. If everyone used Face ID, and that data made it into the wild, identity theft would take on a whole new meaning.

Apple said the iPhone X is the basis for the next decade of smartphone technology. It makes you wonder what's next in terms of biometric identification. Instant DNA matching? No lab needed – done in your hand and in real time.

# # #

CYBERBULLYING

Bullying moves online, with dreadful consequences

September 22, 2017

Bullying has been a problem for as long as there have been children in school. In the real world, it still happens with unfortunate frequency. The effects of bullying can be immediate in terms of changed social behavior and poor academic performance. The effects can also be long-term, profoundly affecting impressionable youngsters for the rest of their lives. Sadly, bullying has now crossed into the online world as well, in the form of cyberbullying.

According to the www.stopbullying.gov website, cyberbullying is "bullying that takes place using electronic technology, such as cell phones, computers, and tablets as well as communication tools including social media sites, text messages, chat, and websites. Examples of cyberbullying include mean text messages or emails, rumors sent by email or posted on social networking sites, and embarrassing pictures, videos, websites, or fake profiles."

At first glance, it would seem that cyberbullying is not as detrimental as actual bullying. After all, there are no physical altercations. No one is going to steal your lunch money through threats of violence. But cyberbullying brings its own set of problems, many of which are unique due to the use of technology by the bullies that can make life for the bullied miserable 24/7, not just at school. It's a case where the old saying, "sticks and stones may break my bones, but names can never hurt me" is demonstrably untrue. Things are different in cyberspace.

Someone can be cyberbullied no matter where they are: at school or at home, when alone or with others, at any time of the day or night. This gives the bullies longer reach than they used to have, making it nearly impossible for the bullied to escape. The bullying is always in front of them.

The Internet makes it possible for bullies to remain anonymous. All you have to do is read the venomous posts on public forums to experience the rancor that even adults exhibit when online. Hurtful messages and embarrassing photos or videos can be posted online and widely distributed. Once something is online, it never disappears. Even deleted files have copies that remain in someone's possession, just a click away from being reposted and restarting the cycle of grief. Cyberspace never forgets.

# # #

STOPPING CYBERBULLYING

Technology and education can help

September 29, 2017

October is national "Stop Bullying" month. The website www.stopbullying.gov has a lot of very useful resources for parents and educators to learn more about bullying and cyberbullying. If you suspect that your child or student is being bullied online, there are steps you can take to address the problem.

The first step is to prevent cyberbullying from happening in the first place. The U.S. Department of Health & Human Services (HHS) suggests the following. Be aware of what your kids are doing online. You should know what websites your kids visit, what apps they use on their smartphones, what social media they use and who their "friends" are. If you suspect something is going on, you can install monitoring software and enable parental control filters on their computing devices. Be prepared for some pushback, but rest assured that you are doing this for their benefit, not just to be nosy.

HHS also suggests establishing rules for technology use by your kids, and educating them about the implications of improper online behavior. Remind them not to share their passwords with their friends. Talk about what websites they can visit, and which ones they cannot (and tell them why). Most importantly, stress that they need to be extremely careful when they post anything online, such as messages or photos. They may think the information can be deleted later, or apps like Snapchat will cause the information to disappear after a few seconds, but nothing ever truly goes away on the Internet. Once it's out there, it's out there forever.

Lastly, create an atmosphere of trust with your kids, so that if they are being cyberbullied, they are not afraid of letting you know. Schools typically have policies in place to deal with bullying. There is technology available to identify the originator of malicious postings; even when the cyberbully thinks they are anonymous, they usually are not. If law enforcement gets involved, their ability to lift the veil of online secrecy is even more potent.

Cyberbullying is such an important issue that the White House has become directly involved. The First Lady has started an anti-cyberbullying initiative to raise awareness of the problem. Sadly, we may have to accept that cyberbullying is the new normal.

# # #

GRAVITATIONAL WAVES

Nobel Prize from Einstein in 1915 to LIGO in 2015

October 6, 2017

The 2017 Nobel Prize in Physics was awarded to three American scientists for their work in detecting gravitational waves: Rainer Weiss, Barry Barish, and Kip Thorne. Weiss is with the Massachusetts Institute of Technology, while both Barish and Thorne are with the California Institute of Technology. All hold the position of professor emeritus at their respective institutions.

The Nobel Prize comes with a cash award of around $1.1 million. The award will be split, with half going to Weiss and the other half shared between Barish and Thorne. The official ceremony takes place in Sweden in December.

Gravitational waves were first proposed by Albert Einstein in 1915, as part of his general theory of relativity. According to NASA, Einstein believed that gravitational waves would be formed when two massive bodies, such as stars, orbit one another or collide into one another. The interaction would create ripples in the fabric of space-time, rather like dropping a pebble in a pond. The waves are invisible, travel at the speed of light, and can squeeze and stretch anything in their path. However, Einstein knew that these physical changes would be so small that he doubted we'd ever be able to detect them.

Weiss, Barish, and Thorne designed a sophisticated apparatus, the Laser Interferometer Gravitational-Wave Observatory (LIGO), to detect gravitational waves. LIGO consists of two L-shaped detectors, each 2.5 miles long and identical in construction. One is in Hanford,

Wash., and the other is across the country, in Livingston, La. Minute perturbations in LIGO's structure caused by gravitational waves are measured by a system of mirrors and lasers. LIGO must account for other forms of movement, such as earthquakes, and eliminate them from the experiment.

On September 14, 2015, almost exactly one century after Einstein's theoretical prediction, LIGO registered the presence of gravitational waves. These ripples in space-time were 10,000 times smaller than the width of an atom. The waves were produced by a massively-powerful collision of two far-flung black holes that occurred 1.3 billion years ago. LIGO has since detected three more instances of gravitational waves, proving that the first event wasn't erroneous.

The detection of gravitational waves was an incredible achievement. It relied on theoretical physics as the basis for the experiment but leveraged new technological and engineering developments to realize the dream. It demonstrates the value of funding basic scientific research, even in times of fiscal restraint. The results from LIGO offer insight into the very nature of the universe, including new theories about the cosmos shortly after the Big Bang.

It might have taken us a hundred years to get here, but we're just getting started.

# # #

TECHNOLOGY ON THE SPACE COAST

Sunny weather, warm oceans, and rocket launches

October 13, 2017

When I first visited the Space Coast in 2002, I was pleasantly surprised to find a unique feature of the beachside hotels. They always had an information board outside the bar, just before you ventured onto the beach. The board displayed the weather forecast, the ocean temperature, and the next launch time. It was the last addition that really impressed me: nowhere else but the Space Coast would rocket schedules play such an important role in daily life. I was sold.

Nearly fifteen years later, I'm still amazed at the role technology plays in this community. There are very few areas that have such a high concentration of skilled labor. I only know of cities like Albuquerque that have a workforce with a comparable percentage of advanced degrees. In New Mexico, that's primarily due to the presence of federal laboratories like Los Alamos. For us, it continues to be the space center and all the aerospace and defense companies that are here to support it.

But this is not your grandfather's space center. Recent years have seen rapid changes in the space program, with a proliferation of private industries like SpaceX taking center stage. NASA is still a huge influence, but there's clearly momentum behind commercial space initiatives. This momentum carries with it a myriad of new technologies in areas such as advanced manufacturing, drones and robotics, and launch systems.

This Friday, October 13, 2017, the Space Coast Tech Council (SCTC) and the Melbourne Regional Chamber are hosting Techxpo at the Eau Gallie Civic Center. It's an excellent opportunity to experience some of the fabulous technology-based products and services coming out of the Space Coast. Registration is just $10. (It's free for SCTC members, and for students and educators with a valid ID). More information about this event can be found at http://bit.ly/2ydvedD.

Techxpo also includes several breakout sessions, which offer the opportunity to hear industry experts discuss topics of interest. This year there are sessions on cybersecurity (which I am moderating), drones & robotics, and aerospace. The cybersecurity session is particularly timely, given the recent data breaches at companies such as Equifax and the loss of highly-classified NSA data through the use of Russian antivirus software. Cybersecurity is something that all organizations need to take very seriously – and there's obviously a lot of room for improvement across the board.

The shuttle program may be gone, but there are still plenty of launches to keep us impressed. This week's SpaceX launch of a used Falcon 9, and the subsequent landing of the booster, are testament to the continued success of technology on the Space Coast. I'm still sold.

# # #

BLADE RUNNER 2049

Is our future one of replicants and holograms?

October 20, 2017

I recently watched the new movie "Blade Runner 2049," the sequel to the iconic masterpiece "Blade Runner" from 1982. I found "Blade Runner 2049" to be an excellent film, full of thought-provoking ideas about advancements in technology and what it means to be human. I also found it to be a rather grim view of our possible future.

In the original "Blade Runner," Deckard (Harrison Ford) is tasked with killing four replicants who have escaped from off-world colonies and returned to Earth, ostensibly to meet their maker. Replicants are bioengineered synthetic humans, designed and created by Tyrell Corporation, to work as slaves. They are hard-wired to die after four years. Deckard begins to question the righteousness of his mission when he meets a replicant named Rachael. Rachael is initially unaware that she's a replicant because her memories seem so real to her, but they are just implants from someone else.

"Blade Runner" was based on the 1968 novel "Do Androids Dream of Electric Sheep?" by Philip K. Dick. It was set in a neo-noir version of Los Angeles circa 2019, where it's raining all the time. There's a distinctly Asian feel to the setting. The technology used for the replicants is impressive, but not too far advanced that we can't relate to it today.

In contrast, "Blade Runner 2049" is set an additional 30 years in the future. Much of the story takes place in a desolated Las Vegas, where there are hints of a past nuclear war and widespread data loss.

The cinematography depicts the desert as perpetually swathed in yellow dust and a damaged environment. The atmosphere is markedly different from the darkness of "Blade Runner," but it's no less depressing. This feeling is enhanced by the harsh industrial soundtrack by Hans Zimmer, which is in stark contrast with Vangelis' quiet and ethereal score for the original film.

What really struck me was the technological advancements that have been made in 30 years. They seem both magical and colder. The time shift into the future brings us into a world so alien that it reminded me of the time traveler from H.G. Well's "The Time Machine," who goes 800,000 years into the future and barely recognizes the landscape.

In 2049, computing has become ubiquitous through implants and unseen networks. Replicants mingle with humans, but they are still shunned as "skin jobs" and treated poorly. The Blade Runner called K (Ryan Gosling) knows he's a replicant, but he's been programmed to perform his duties without remorse.

It's remarkable that a central relationship in "Blade Runner 2049" is between two artificial beings: a replicant and a hologram. Wither humanity?

# # #

NET NEUTRALITY

Is open access threatened by changes in Internet legislation?

October 27, 2017

This is Open Access week. Open access is a movement to provide virtually unrestricted access to various forms of intellectual property. For example, in the traditional academic publication model, the author does not pay for their work to be published in a journal, but readers who want access to the published work pay the publisher certain fees. In an open access publication model, the roles are reversed: the author pays a processing charge to the publisher, but the content is made freely available to everyone online.

Advocates of the open access model cite the success of similar sharing models, such as the open source movement in the software community, as evidence for the viability of this new way of communicating results. This is particularly true for research that has been funded using public money, such as grants from the National Science Foundation or the National Institutes of Health.

Open access even applies to textbooks. The author is paid a stipend to produce a monograph on a particular topic, but the rights to the book are placed in the public domain, and there are no royalties. As long as quality control measures are in place, this publication model can also work. However, as someone who makes part of my living from my writing, I have to admit that I'm a bit conflicted about this development. I understand the concept that "information just wants to be free," but I also like getting paid for my work.

Irrespective of the compensation model, I firmly believe information on the Internet should be available to everyone. "Access for All" was a mantra of the early World Wide Web. Your broadband service provider should not be able to block certain websites, deny the use of select apps on your smartphone, or throttle your download speeds for specific content. These tenets are at the center of the net neutrality movement, which is again making headlines.

In 2015, the Obama administration placed Internet access under an old statute called Title II from 1934. This gave the federal government more oversight and regulatory powers over Internet access. The current administration may soon reverse this ruling. The tech companies are lined up (mostly) on one side, with the large carriers lined up on the other side. It's a complicated topic, but one that may impact open access – and impact you too. Stay tuned!

# # #

SEVENTH ANNIVERSARY

Seven years of Technology Today

November 3, 2017

I've been writing the "Technology Today" column for a little over seven years. My first column, "Life with Kindle: So far, so good," was published on Oct. 23, 2010. Interestingly, it's the tenth anniversary of the release of the first Kindle e-book reader, and it's still going strong. I continue to use it to read the morning news from foreign sources, but I also still enjoy the printed version of the newspaper with breakfast.

Seven years is a long time in the technology world. In thinking about what have been some of the most important changes in the role of technology in society, there are plenty of candidates. I think three significant developments have been the role of social media in shaping public perception, the continuing challenges of cybersecurity, and the rise of big data and machine learning. All three had substantial influence on the last presidential election.

Social Media: According to statista.com, Facebook had 2.07 billion monthly active users as of the third quarter of 2017. When I started this column, they had just over 600 million users. Together, Facebook and Google act as a duopoly, controlling the vast majority of digital advertising. No one else even comes close.

The power social media companies now enjoy has gone far beyond sharing cat pictures. They are able to change how people think about the news and the world around them. With Congress currently looking into possible Russian influence in the 2016 presidential election through offshore advertising and the use of online bots, it's clear that social media is now a big part of the mainstream media.

Cybersecurity: A few years ago, I was optimistic that we'd get better at cybersecurity. Now, I think things are getting worse. Software permeates almost every aspect of our lives, and the programs are increasingly complicated. They are networked into everything, and the Internet-of-Things opens even more doors for hackers to enter.

Rarely a day goes by without a news story about a major data breach. We're on the defensive as nation-states attack our public infrastructure constantly. This is also something Congress is looking into vis-à-vis the 2016 election.

Big Data and Machine Learning: We are awash in data. All of our devices throw off data like a dog shakes off water. This data is scooped up by sensors and used as input to sophisticated machine learning algorithms.

These algorithms perform predictive analytics – determine future outcomes based on past events and current trends. But we all know how that worked out during the last president election. Groupthink and incorrect data produced erroneous results.

Who knows what will happen in 2020. More fake news?

# # #

PRESIDENT #TWEET

What havoc will an extra 140 characters creak?

November 10, 2017

Christmas has come early to tweeters everywhere! Twitter has doubled the length of tweets, from 140 to 280 characters. Apparently, there was a lot of anxiety among heavy Twitter users (primarily in English) about the limitations imposed by the 140-character limit. Instead of big words, they switched to emojis to convey meaning. Now, they are free again to update the world about their latest activities in even greater detail.

According to statista.com, Twitter currently has 330 million monthly active users. This is a far cry from Facebook's 2 billion users, but the two social media services arguably serve different purposes. Twitter has become the bullhorn for those who want to be heard across the cacophony of the Internet. Probably no one makes better use of Twitter than President Trump.

According to Twitter, @realDonaldTrump joined the microblogging service in March 2009. Since then, he's been quite prodigious, sending out over 36,000 tweets. He has over 42 million followers, but he follows just 45 people. I wonder what this massive difference in speaking versus listening says about a person.

I admit I do tweet, but only occasionally. For example, I send out a tweet when this column is published, but that's about it. I joined Twitter in March 2011, but in contrast to President Trump, since then I've only sent out 381 tweets. I probably should do more to build my brand, but I don't.

There have been several interesting writing projects on Twitter. Shakespeare's plays have been published as a series of tweets. Entire flash mob novels have been published on Twitter. Short poems seem particularly popular when turned into tweets.

I have no doubt that sending short messages and tweets has affected how we communicate. For many people today, lengthy emails are just too onerous to plow through. If you can't convey your message as a tweet, don't bother. This seems in agreement with our shrinking attention spans, but I'm not sure which is the chicken and which is the egg in this case.

President Trump uses Twitter to bypass the media and reach the population directly. Some people think he's reckless with his tweets, while others think he's an effective communicator in an age of information overload and fake news. In his new book, "Winning Bigly: Persuasion in a World Where Facts Don't Matter," Dilbert creator Scott Adams argues that Twitter is just part of the President's strategic approach to persuasive information dissemination. I wonder what he will do with all the extra space.

There are 1,457 characters (counting spaces) in the Gettysburg Address. In a little more than five tweets, Lincoln would be done. Isn't progress marvelous?

#

SUPERCOMPUTERS

Mind-bogglingly fast and very power-hungry

November17, 2017

Each year, the top 500 supercomputers in the world are ranked according to raw computing power, which is measured in "flops." A "flop" is a floating-point operation, such as multiplying two real numbers. For the first time ever, two Chinese machines sit atop this lucrative list; the US continues to lose ground to Asia in this important race.

The fastest computer is the Sunway TaihuLight, which is installed in the National Supercomputing Center in Wuxi, a town not too far from Shanghai. The computer has been clocked at 93 petaflops. A "peta" is 1 quadrillion, which is 1,000 trillion. Take a moment and think about how fast that is: the TaihuLight can perform 93 million billion calculations per second. TaihuLight has 10,649,600 cores, or processors. For comparison, my MacBook Air has 2 cores.

The No. 2 machine is the Tianhe-2, which is installed at the National Supercomputer Center in Guangzho. It scored 33.86 petaflops – still exceedingly fast, but only about a third as fast at the TaihuLight.

According to top500.org, machines 5-8 are US-based supercomputers. No. 5 is Titan, a Cray XK7 running at 17.5 petaflops at the Oak Ridge National Laboratory. No. 6 is Sequoia, an IBM BlueGene/Q running at 17 petaflops at Lawrence Livermore National Laboratory. No. 7 is Trinity, a Cray XC40 running at 14 petaflops at Los Alamos National Laboratory. No. 8 is Cori, also a

Cray XC40 running at 14 petaflops at Lawrence Berkeley National Laboratory.

It's no coincidence that all these machines are housed in national labs. The supercomputers are used for complex tasks such as simulating nuclear weapons blasts, modeling global climate patterns, and cryptography. They are also tremendously expensive. For example, Titan cost nearly $100 million. But this pales in comparison to the Japanese machine called K, the No. 10 ranked computer, which runs at 10.5 petaflops and is made by Fujitsu in Japan. K reportedly cost $1.25 billion, making it the most expensive computer ever built.

According to cnet.com, an IBM-built machine named Summit at the Oak Ridge National Laboratory may help the US leapfrog back into the top of the rankings. Summit is designed to reach 200 petaflops. The timeline to reach this milestone is unclear.

The power consumption of these machines is incredible. For example, TaihuLight uses 15 megawatts; Summit will need a 20-megawatt supply. That's enough juice to power over 16,000 homes.

The next major supercomputing goal is to become the first exascale machine, running at 1,000 petaflops: a billion billion calculations per second. Heaven knows what the power requirements for such a behemoth would be, but I sure wouldn't want to pay the FPL bill.

###

STAR WARS TECHNOLOGY

Some of the Force is already with us

November 24, 2017

It seems like everyone is gearing up for the upcoming release of the next installment in the "Star Wars" franchise: "Star Wars: Episode VIII – The Last Jedi." The plot is summarized online as, "Rey develops her newly discovered abilities with the guidance of Luke Skywalker, who is unsettled by the strength of her powers. Meanwhile, the Resistance prepares to do battle with the First Order."

It all sounds very exciting, but what's really interesting to me is how the technology in "Star Wars" has influenced society here on Earth. Remember, we're told "Star Wars" took place "a long time ago." It's one of the few science fiction stories that is set in the past, not in the future.

AI and Robots: The two best-known robots from "Star Wars" are the trashcan-shaped droid R2-D2 and the human-form protocol droid C-3PO. Both are imbued with artificial intelligence, but the droids are very different. For example, R2-D2 doesn't speak as much as it bleeps and chirps. However, these machine noises clearly represent a language, because C-3PO understands it, as do some characters like the young Anakin Skywalker who originally built C-3PO.

In the multi-cultural "Star Wars" universe, language skills are particularly fascinating. Characters like Han Solo understand many languages and dialects, as witnessed by his interactions in the Cantina bar with numerous aliens. But he doesn't seem to be able to speak any language other than English, which the aliens appear to understand. Perhaps everyone used the equivalent of Babel Fish in the old Republic.

Holograms: One of the most famous scenes from the original "Star Wars" movie, 1977's "Star Wars: Episode IV – A New Hope," is the message from Princess Leia to Obi Wan Kenobi. Leia pleads for help from the old Jedi master, saying, "Help me, Obi-Wan Kenobi. You're my only hope." The message is stored in R2-D2 and recorded as a hologram.

Who needs text messages or voice mail when 3D projections are available? Personally, I'd jump all over holograms as the next "You've got mail." Think of all the typing it would save!

###

STAR WARS SOCIETY

Absolute power corrupts absolutely

December 1, 2017

The creator of "Star Wars," George Lucas, is quoted as saying, "Power corrupts, and when you're in charge, you start doing things that you think are right, but they're actually not." He was probably referring to Republic (which later became the Galactic Empire), which was misled by their elected leader, Chancellor Palpatine, who later turned out to be the evil Sith Lord, Darth Sidious. But Lucas may also have been referring to the Jedi Council, a group of unelected Jedi Masters who attempted to guide the Republic from behind the scenes.

As the Republic morphs into the Empire, the Senate grants Palpatine increasingly far-reaching emergency powers to address the rebellion by the separatists. Quite soon, personal freedom is severely limited, and Palpatine becomes the first unelected Emperor of the galaxy. He maintains his grip on power through the use of a clone army and the politics of ultra-nationalism.

Technology plays an important role in "Star Wars" society, for both good and bad. For example, the Rebellion relies on small spacecraft, such as the X-Wing fighter, to engage the forces of the Empire. But the Empire has vastly larger ships, such as Star Destroyers, that are packed with much more weaponry than the rebels can ever match. The Empire also has massive Death Stars, which get larger with each episode. They are eventually destroyed, but at great cost.

It seems that most of the Empire's resources are directed at building up their military. Education is rarely mentioned in "Star Wars" movies, with a few exceptions. Younglings are shown being trained to become Jedi, but they are few in number relative to the galactic population, and they are slaughtered by Anakin Skywalker in a fit of rage while following the orders of the Emperor. Clone soldiers are shown attending an education facility, but the goal is to mature them more rapidly than normal, not to better their lives. Poor education ultimately leads to poor job prospects for members of the Empire: serf, soldier, or slave, with a few entrepreneurs trying to eke out a living.

This is fiction of course, but the echoes of our own history, past and present, are clear. Consider a certain "Rocket Man" in his hermit kingdom today. Nuclear ambitions and a starving population with little hope for a better future. Be grateful that's not us. Yet.

# # #

CSEdWeek

Choose a career in computing

December 8, 2017

According to the official website (https://csedweek.org/about), "Computer Science Education Week (CSEdWeek) is an annual program dedicated to inspiring K-12 students to take interest in computer science." This year, CSEdWeek runs from December 4-10. You may have noticed several "Hour of Code" events happening around the community.

Encouraging young people to consider careers in a STEM-related field is an ongoing effort, but one that's incredibly important to our national interest. Computing in particular has been singled-out by the president as the primary area of need. The bottom line is that we need more computer scientists, and to meet this need, we need more students to enter the computing field.

My own path to a career in computing was not direct. In high school, I was primarily interested in electronics. I spent countless hours in the lab doing TV repair (which later served me well for summer jobs), designing and building printed circuit boards, and generally getting my hands dirty. But I loved every minute of it.

When I began my undergraduate studies, I was in the electrical engineering program. My exposure to computers was rather limited at that point, so I just assumed EE was where I wanted to be. But my first programming course changed all that. I quickly realized my passion was not with the components, but with the computer made from the components. I had to take extra courses over the summer to make up for lost time, but I switched majors to computer science at the start of my second year.

I chose to get a master's degree in computer science in part because I knew there was so much more I could learn. I also knew that getting a graduate degree in computing would greatly increase my job prospects (which it did). I had already re-entered the workforce and was enjoying considerable success when the academic bug bit me again, and I chose to pursue a Ph.D. Getting a Ph.D. is not for everyone, but it opens many doors that remain closed otherwise – such as being a professor.

The very first Ph.D. degree in computer science was awarded to Richard Wexelblat in December 1965 at the University of Pennsylvania. He was the first person to have the words "computer science" explicitly on his doctoral credentials. Nearly 30 years later, in January 1995, I earned my Ph.D. in computer science. Obtaining this degree was one the best decisions of my life. I can't think of any field other than computing that offers such a varied career, full of interesting and rapidly-changing topics.

For anyone considering their own career in computing, I say "Go for it!"

# # #

AOL AIM Shuts Down

Goodbye to the original social media program

December 15, 2017

After 20 years of service, AOL's instant messenger, AIM, is shutting down. Effective December 15, 2017, AIM will be digitally deceased. With its passing goes one of our last connections to the origins of social media.

When AIM was released in May 1997, it was a very different world. Windows 95 was the most popular operating system. Most people relied on dial-up modems to connect to the Internet. The iconic phrase "You've got mail," the audio reminder from AOL's email program, became synonymous with being connected to the fledgling online world.

AIM was not the first chat program, but it was the most widely used. According to MIT's Technology Review, "At its peak in 2001, AIM had 36 million active users; as of this summer, it had just 500,000 unique visitors a month." That's a tremendous drop-off in users, but it's reflective of a technology that failed to keep up with the times. With its proprietary OSCAR and TOC protocols, AIM just couldn't compete with text messaging on smartphones, tweets on Twitter, or with apps like WhatsApp, Facebook Messenger, and Instagram. AOL does not plan a replacement for AIM; it's just going dark.

Computing history is unique in that "back in the day" can literally mean just a few days ago. In AIM's case, it's history spans two decades. But in that short period of time, AIM was responsible for introducing the world to message abbreviations like "LOL" and emoticons like smiley faces. Some language purists lament the use of these shorthand notations, claiming that they are ruining our communication skills. I think there's some truth to that, but I also think it's too late to worry about it. An entire generation of youngsters have now grown up more comfortable texting than speaking, and you can thank (or blame) AIM for starting the process.

AIM connected users in real-time in ways that had never happened before. For the first time, you could see if your friends were online – anywhere in the world. It was a transformative experience for many teenagers. Facebook Friends are just a modern version of this capability.

If you've been a heavy AIM user, and you want to capture your message history, act fast. Very soon, your data will be gone. So will your Buddy List. I never really got hooked on AIM, so this is not a problem for me. But someday in the not-too-distant future, I could find myself scrambling to save my own communication history from programs like Apple's iMessage. It's a stark reminder that apps come and go, but data is our most valuable commodity. Save your screen name!

# # #

HOLIDAY TRAVEL

Bring back the Concorde

December 22, 2017

We've hit the peak of the holiday season, which means millions of travelers will be taking planes, trains, and automobiles in a frantic rush to get somewhere. If you've had the misfortune of experiencing any of these modes of transportation in our major cities recently, you know any last drops of pleasure have long since been squeezed out of modern travel. It's more of a stressful endurance test than anything else.

The unprecedented power outage at Hartsfield–Jackson Atlanta International Airport over the weekend is still without a satisfactory explanation. How can the world's busiest airport be so susceptible to a single point of failure? I thought we built airports to be resilient, but it seems ATL is more like a shaky house of cards.

The inaugural trip for a new high-speed train in Washington State ended in disaster. The locomotive was reported to be traveling 80 mph in a 30 mph zone. There are control systems that can automatically slow down trains that are out of control, but the system wasn't installed. The driver may have been distracted, but if that's all it takes to cause a derailment, we've not made much improvement in train travel in a hundred years.

Our clogged roadways are not much better. Even taking bus tours is fraught with danger. Over a dozen people were killed when a bus carrying tourists from cruise ships flipped over while driving to Mayan ruins in Mexico. The cause of the accident is still being investigated. And speaking of cruising, twice recently, hundreds of passengers have become sick at sea with stomach illnesses.

What's the solution to our travel woes? I've been thinking about this a lot lately as I prepare to jet off towards the Great White North for the holidays. I think the answer is to limit the amount of time actually spent in transit. Unless we can engineer Star Trek-like transporter systems, the best we can do is bring back supersonic air travel.

It's been over 50 years since the Concorde prototype was unveiled to crowds in Toulouse, France. The closest I ever got to flying on these needle-nosed wonders was seeing one on the tarmac at JFK. Traveling at Mach 2 while sipping champagne on the 3.5-hour flight from London to New York sounds a lot more attractive than sitting in the middle seat of a crowded, long haul, no frills bus in the air.

The Concorde stopped flying in 2003 in part because it was not economically viable. But maybe focusing new technology on an updated Concorde would be a better use of our expertise than rolling out the latest flatbed seats.

# # #

Looking Back at 2017

Bitcoins, breaches, and busts

December 29, 2017

Another fast-paced year in technology is coming to an end. Looking back at 2017, we celebrated the 40th anniversary of Star Wars and the Voyager probes, net neutrality was repealed, and the latest iPhone was introduced with Face ID. But to me, there were three things that really stood out: bitcoins, breaches, and busts.

Bitcoins: If you were fortunate enough to have bought a bitcoin at the start of the year, you would have realized a paper gain of over 1,500%. Assuming, that is, that you were able to sell the bitcoin before its value crashed back to Earth just before the holidays. I think investing in bitcoins is more akin to gambling in Las Vegas, and in both cases, the house always holds the winning hand.

Bitcoins are a form of cryptocurrency and just one implementation of blockchain, a technology for a global, distributed ledger system that many people think represents the future of commerce. The major investment houses started a derivatives market for bitcoins in December, which made an already volatile situation even more so. Time will tell what effect bitcoin, Ethereum, Litecoin, Ripple, or any other digital money will have on the economy.

Breaches: Data breaches were in the news nearly every week all year. As more computing moves into the cloud, the data follows it. Poor programming practices, pervasive human error, and sloppy organizational governance all contributed to this problem. As in previous years, breaches involved a lot of personal data, but for the first time, that fact that parts of our national infrastructure were hacked was made public. Leaks from government agencies around the world occurred at an alarming rate.

Cybersecurity is a tough nut to crack. But it seems we lack even cheap party nutcrackers, never mind big hammers. At some point, we're going to have to change the rules of the game, because at the moment, we're losing.

Busts: There's always been regrettable instances of poor behavior in the workplace, but 2017 was unique in the number of public busts that took place. Alleged perpetrators were outed through the power of social media, something that would not have been possible even a few years ago. The sheer number of reports related to well-known personalities in entertainment, media, and finance was astounding.

Many of the alleged events took place many years ago, which almost always leads to "he said / she said" arguments that are difficult to resolve. But recordings, especially those that are placed online, never go away. They offer objective evidence that is difficult to refute. Sexism in technology is rife, but perhaps technology can help rectify the problem in 2018.

# # #

ABOUT THE AUTHOR

Scott Tilley is a professor at the Florida Institute of Technology, president and founder of the Center for Technology & Society, president and co-founder of Big Data Florida, president of the Space Coast chapter of INCOSE, and a Space Coast Writers' Guild Fellow. His recent books include *Holidays* (Anthology Alliance, 2017), *Surreal Technology* (CTS Press, 2017), and *Systems Analysis & Design* (Cengage, 2016). He writes the weekly "Technology Today" column for *Florida Today*. More information about all of his writing is available online at http://www.amazon.com/author/stilley.